WHAT'S INSIDE THE PROTON

THE INVISIBLY OBVIOUS

Written by: Ofer Comay

Scientific advisor: Eliyahu Comay

Editor: Addie Ney

Copyright Notice

"Even now, two decades after QCD was formulated, little is known from first principles about the structure of the proton, neutron and other hadrons."

Robert L. Jaffe, 1995[1]

"The proton is complicated, but it is a very, very important object in our lives. It is unsatisfying intellectually that we cannot understand how the inside of the proton behaves."

Emlyn W. Hughes, 1997, about the proton spin crisis[2]

"If the results are not a statistical fluke, new physics has been observed. One possibility is that our understanding of what's inside the proton is somehow wrong."[3]

Frank Sciulli , 1997, about DESY experiments

"The results are in complete disagreement with the calculations... We are not aware of any published detailed prediction presently available which can explain the behaviour of these data."[4]

J.J. Aubert et al., 1983, about the 1st EMC effect

---

[1] Physics Today, 1995

[2] Science News, Sept 6, 1997 by Ivars Peterson

[3] Columbia Unversity News, 1997

[4] J.J. Aubert *et al.*, Phys. Lett. **123B**, 275 (1983)

"Ironically, from the perspective of QCD, the foundations of nuclear physics appear distinctly unsound."

Frank Wilczek, 2007, about QCD vs. the strong nuclear force[5]

"Currently, the color van der Waals force does not seem to be a correct model for nuclear interaction without modifications."

S. S. M. Wong, 1998, about QCD vs. the strong nuclear force[6]

"No direct translation between the Standard Model and VMD has yet been made."

H.B. O'Connell, B.C. Pearce, A.W. Thomas and A.G. Williams, 1997, About the hadronic properties of the photon[7]

"The whole story – the discoveries themselves, the tidal wave of papers by theorists and phenomenologists that followed, and the eventual "undiscovery" – is a curious episode in the history of science."

C.G. Wohl (LBNL), a review about the search after the Pentaquarks, 2008[8]

---

[5] Frank Wilczek, *Hard-core revelations*, NATURE, Vol. **445** 156 (2007).

[6] S. S. M. Wong, *Introductory Nuclear Physics*, (Wiley, New York, 1998). p.102

[7] H.B. O'Connell, B.C. Pearce, A.W. Thomas and A.G. Williams, *Rho-omega mixing, vector meson dominance and the pion form-factor*, Prog. Nucl. Part. Phys. 39 (1997) 201-252

[8] C.G. Wohl (LBNL), *Pentaquarks*, 2008. pdg.lbl.gov/2009/reviews/ rpp2009-rev-pentaquarks.pdf

"No strangelets were found in the experiment."

K. Han, a report about the search for stable Strange Quark Matter
in lunar soil[9]

*"If you see nothing, in some sense then, we theorists have been
talking rubbish for the last 35 years."*

John Ellis, about the search for the Higgs Boson in LHC, 2007[10]

*"There are many students who have never seen data; I don't know
how much longer we can keep going like that."*

Michelangelo Mangano, about LHC experiment, 2007[11]

"It will probably be the end of particle physics."

Martinus Veltman, about a scenario in which the Higgs fails to show
up at the LHC, 2008[12]

---

[9] K. Han et al., Phys. Rev. Lett. 103, 092302 (2009)

[10] The New York Times, 2007, www.nytimes.com/2007/05/15/science/15cern.html

[11] Ibid.

[12] www.lindau-nobel.org/upload/Web_As_LHC_Draws_Nigh__Nobelists_Outline_Dreams__And_Nightmares__02_07_1474.pdf

TABLE OF CONTENTS

PREFACE

About this Book

Eliyahu Comay is a theoretical physicist who has published scientific articles in the areas of particle physics, quantum mechanics, nuclear physics and electromagnetism.

Many of his articles discuss fundamental issues. In 1987 Comay showed that the electric Aharonov-Bohm effect cannot occur since it violates energy conservation.[13] Indeed, this part of the effect predicted by Aharonov and Bohm in 1959,[14] was not discovered.[15,16,17] In another issue, Comay provided a consistent explanation[18] to the "hidden momentum" problem, a paradox presented by Shockley and James about thirty years earlier.[19]

---

[13] E. Comay, *Further comments on the original derivation of the electric Aharonov-Bohm effect*, Physics Letters **A120** 196, (1987).

[14] Y. Aharonov and D. Bohm, *Significance of electromagnetic potentials in quantum theory*, Physical Review **115**, 485–491 (1959).

[15] E.g.: "...there isn't any direct experimental observation of the electric AB effect..." A.V. Ghazaryan, K. Moulopoulos, A. P. Djotyan and A. A. Kirakosyan, *Investigation of the electric Aharonov-Bohm effect in a quantum ring,* 50 years of the Aharonov-Bohm effect, Tel-Aviv University, 2009.

[16] "The existence of electric Aharonov-Bohm effect, that has not been confirmed experimentally, is a very controversial issue." *The Electric Aharonov-Bohm Effect,* Ricardo Weder, 2010. arxiv.org/abs/1006.1385

[17] Batelaan, A.; Tonomura, A. (Sept. 2009). *The Aharonov-Bohm effects: Variations on a Subtle Theme*, Physics Today: 38-43.

3

Comay's main discoveries were not accepted by mainstream physicists. During the last thirty years he published many scientific articles which establish a new model of the strong interaction. His model contradicts the currently accepted model, called quantum chromodynamics (QCD). This book brings the main points of the model in a language that can be understood by particle physicists as well as readers who understand physics at the level of popular science.

Physicists in the 21st century have good reasons why they refrain from considering new theories that contradict apparently "well established theories." The justification of this approach has two levels:

- There are so many "proofs" of mainstream physical theories that the probability of a mistake is very low.

- It might be very difficult and exhausting to get into the details of a new theory and find whether it is valid or not.

In this book I try an approach that aims to remove these "built-in" obstacles. The first unit of the book, "Puzzling Experimental Results," presents only well-known facts that were published in mainstream physics textbooks and journals during the previous century.

This unit doesn't discuss any new model.

---

p.41: "Thus far such experiments, crucial as they are to the characterization of the AB effects, have remained out of reach. Nor has the pulsed version of the original (Type I) electric AB effect have been performed."

[18] E. Comay, *Exposing "hidden momentum,"* American Journal of Physics, 1028-1034 (1996).

[19] W. Shockley and R.P. James, *"Try Simplest Cases" Discovery of "Hidden Momentum" Forces on "Magnetic Currents,"* Phys. Rev. Letters **18**, 876 (1967).

Particle physicists should be able to read this part within 30-45 minutes. Following this less-than-one-hour-effort, many will no longer consider QCD as a "well established theory." It is not only that this part includes a large set of unexplained experimental results. It is the nature of these well established phenomena that would make almost any reader very skeptical about QCD foundations. I believe that after completing this relatively easy effort, many of you will be able to construct parts of Comay's model by yourselves.

Another set of arguments why any such attack is a-priori futile is the fit of experimental findings to QCD. Such arguments were collected via long interviews with particle physics experts, and by consulting the relevant scientific literature. A response to these arguments is provided in one of the concluding chapters – "And Yet, Why Do Scientists Believe in QCD?"

Particle physicists who wish to dive deeper into the issue may find abstracts of selected articles in the last appendix of this book. My advice is to read first the book in order to have an idea about the context of the articles. Undergraduate students and science lovers may consult the appendix "terminology" for definitions of physical terms that appear inside the book.

Eliyahu Comay published scientific articles with predictions which were almost always against wall to wall consensus of the physicist community. Among others, he predicted that the Higgs particle, Dirac monopoles, glueballs, SQM, pentaquarks, di-barions and the electric Aharanov-Bohm effect would not be found. All his predictions were based on profound scientific papers that he published.

To date, none of his predictions was refuted.

In 2008 the Tel-Aviv University physics department asked him to remove the controversial material from his web site. This book can help you understand the issues and take side in this controversy.

Enjoy reading.

Ofer Comay, Tel Aviv, March 2011

UNIT 1: PUZZLING EXPERIMENTAL RESULTS

Chapter 1: An Invitation to Solve a Mystery

This book discusses the structure of some of the most basic constituents of matter, called hadrons, and one of the basic forces, the strong interaction, that operates at its most elementary level. The goal of this book is to bring this subject to the awareness of as large a public as possible. This is not simple, because issues related to this field have always been inaccessible to the general public, and even to most scientists, including physicists. Most people who are interested in the field lack the capacity to distinguish between a reasonable and an unreasonable theory.

As you read this book, you will realize that a central part of today's physics is seriously challenged.

Physicists might argue that you need to study this material in depth, for at least two years, before you can distinguish between what is true and what is false in this field.

Let us carry out a short test. In this chapter you will become familiar with nearly ten unsolved problems in physics. If you have a good understanding of "high school physics," then even without any previous knowledge of these issues, you will discover on your own the key idea which solves many of these problems.

Sound crazy? Just concentrate while you read through several pages containing somewhat technical descriptions of well-known

9

experimental facts. Most of them are considered to be unsolved mysteries in physics.[20]

Ready? Let's begin.

The basics

The atom has a nucleus, which is surrounded by shells of electrons. Electrons have a negative electric charge and repel each other, while they are attracted to the positively charged protons within the atomic nucleus. This force is called "electromagnetic force."

The force holding molecules (and atoms of noble gas) together inside a liquid droplet is called the "van der Waals force," named for the Dutch physicist. The van der Waals force is weaker than the force that holds electrons in neutral atoms and molecules. Thus, as temperature increases, most liquids evaporate before the molecular structure is broken.

Consider the simplified illustrations below: the term "simplified" means that the particles are illustrated as "balls," which they are not, and the size of the nucleus is greatly exaggerated.[21]

---

[20] The "Terminology" appendix explains the meaning of some of the physical terms used herein. Most subjects described in this book contain reference to their scientific origin.

[21] Quantum mechanics assigns wave function to every particle, and the figure shows the electrons as if they were localized like in the Bohr atomic model.

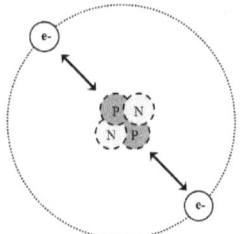 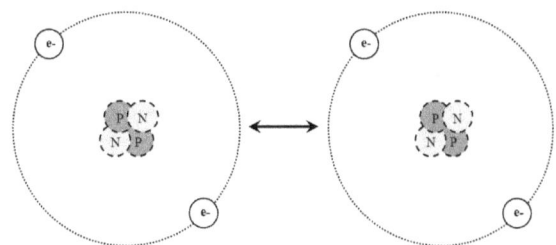

Figure 1. A simplified illustration of the electromagnetic forces operating between electrons and atomic nucleus

Figure 2. A simplified illustration of van der Waals forces operating between two helium atoms. See explanation later.

An experiment conducted at the Stanford Linear Accelerator Center during the late 1960s and early 1970s demonstrated that every nucleon (proton or neutron) is composed of three valence quarks located in its external shell. A significant result of these experiments is that these quarks account only for about one half of the nucleon's mass.[22,23] According to today's dominating theory, quantum chromodynamics (QCD), the other portion of the mass is carried by particles which cannot be directly detected, called "gluons." QCD claims that gluons exist inside the nucleon and "glue" the quarks together. For now, let's leave this claim in question.

The force holding the quarks together is called the "strong interaction."

[22] H. Frauenfelder and E. M. Henley, *Subatomic Physics*, (Prentice Hall, Englewood Cliffs 1991) p.153

[23] D. H. Perkins, *Introduction to High Energy Physics,* (Addison-Wesley, Menlo Park, CA 1987) p. 282

The force holding nucleons inside the atomic nucleus is called the "strong nuclear force."

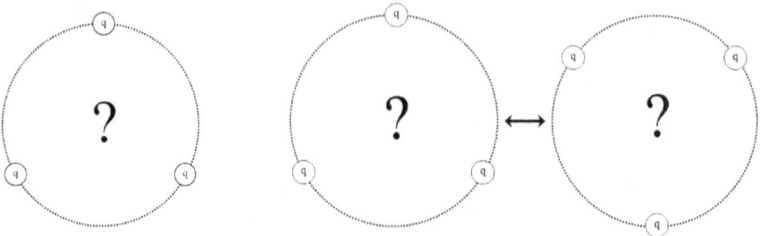

Figure 3. Strong interaction between quarks inside a nucleon

Figure 4. Strong Nuclear Force between two nucleons

The van der Waals force and the strong nuclear force share some interesting features.

Residual forces that vanish at distance

Two of the fundamental forces in physics, the electromagnetic and gravitational forces, operate between bodies, and their intensity decreases gradually as the bodies recede from each other. Unlike these forces, the van der Waals force, acting between neutral molecules, has a particular feature: when the molecules are far apart from each other, the force cancels out rapidly and practically vanishes. It is active only when the molecules are close to each other. This behavior is totally different from the common interaction pattern of the fundamental forces mentioned above.

How does this happen?

Practically, non-ionized atoms and molecules appear to be neutral when measured at a distance. Here, the fields of the positively charged nuclei and the negatively charged electrons cancel each other. This is known as the "screening effect."

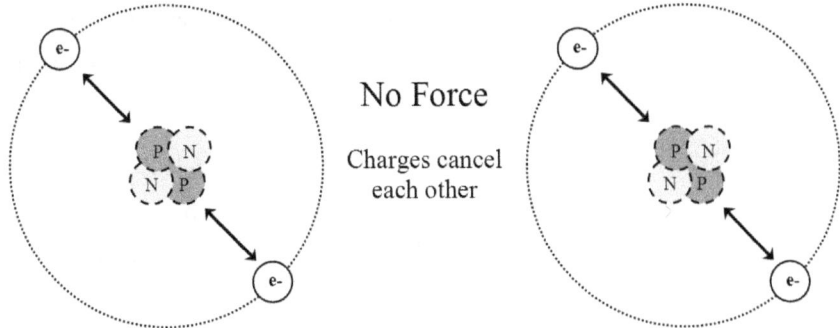

Figure 5. A simplified illustration of the "screening effect" in a pair of helium atoms (containing two electrons each). The nuclear and electronic fields cancel each other when the atoms are not too close.

But when the molecules get close to each other, the electrons in the external shells of one molecule "sense" the electric charge in the neighboring molecule and charge distribution varies. Thus, at appropriate temperatures, the van der Waals force is strong enough to hold the molecular cluster in a liquid state.

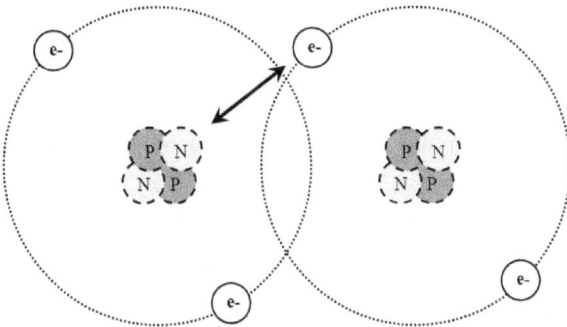

Figure 6. Van der Waals force between pair of 2-electron atoms (He). Electrons of one atom are attracted by the neighboring atomic nucleus.

Similarly, the force holding the nucleons together, the strong nuclear force, is indeed very strong, but it is quite small in comparison to the strong interaction holding the quarks together inside the nucleon.

Furthermore, as the two nucleons move away from one another, the strong nuclear force ceases to apply. This is similar to the behavior of the van der Waals force.

Both the van der Waals force and the strong nuclear force are called "residual forces" in the literature. They are significantly weaker than the fundamental forces from which they derive: the electromagnetic force between nuclei and electrons and the strong interactions between quarks.

Incompressibility

One characteristic of molecules in a liquid state is the familiar phenomenon of being incompressible, which means that a liquid's volume hardly decreases when pressure is applied to it. In fact, a liquid's specific volume[24] is almost constant, because when two molecules move toward one another, they first feel attraction due to the van der Waals force, but at a certain distance, a strong repulsive force appears. (This is due to the Pauli exclusion principle).[25]

A similar phenomenon is the density of nucleons inside the atomic nucleus. The nucleon density in a large nucleus is almost identical to that in a small nucleus (with the exception of very small nuclei).[26]

Distance dependence of the potential

The curve in Figure 7 represents the distance dependence of the potential between two molecules.[27] The steep declining part of the

---

[24] "Specific volume" is the volume per unit of mass.

[25] For an explanation of Pauli's Exclusion Principle, look in the "Terminology" appendix.

[26] S. S. M. Wong, *Introductory Nuclear Physics*, (Wiley, New York 1998) p.139

[27] For a similar graph, see H. Haken and H. C. Wolf, *Molecular Physics and Elements of Quantum Chemistry*, (Springer, Berlin 1995). p.15

curve on the left is attributed to the repulsive force that stems from the Pauli exclusion principle. The lowest point in this curve is the equilibrium state between the van der Waals attraction and the repulsive feature of the Pauli exclusion principle. The increase on the right is attributed to the van der Waals force that vanishes in distance (the curve touches the x-axis).

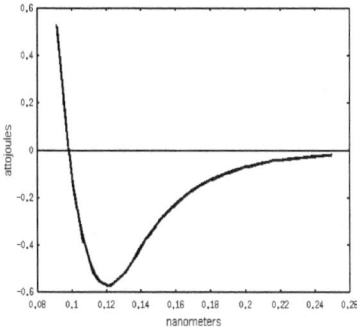

Figure 7. Potential – distance dependence of molecules

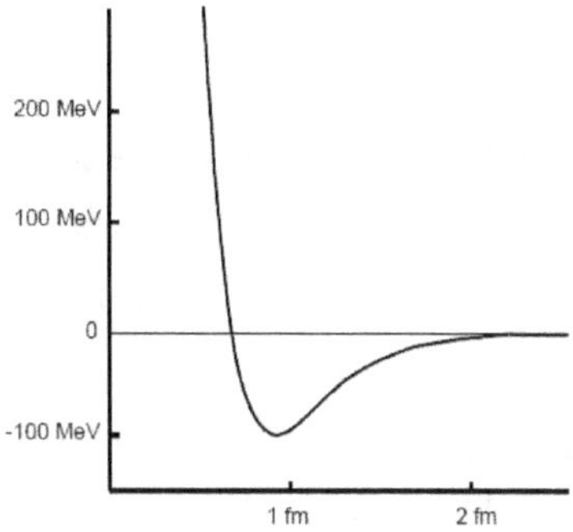

Figure 8. Potential – distance dependence of nucleons

The curve in Figure 8 describes the distance dependence of the strong nuclear force potential.[28] Needless to say, the two curves have very similar characteristics.

The current theoretical status of the strong nuclear force

The similarity between the van der Waals force and the strong nuclear force has been known for at least 70 years. However, nuclear physicists maintain[29] that unlike the van der Waals force, the nuclear force cannot be derived from the current theory of strong interaction.

---

[28] For a similar graph, see S. S. M. Wong, *Introductory Nuclear Physics*, (Wiley, New York 1998). p.97

[29] S. S. M. Wong, *Introductory Nuclear Physics*, (Wiley, New York, 1998). p.102

Even QCD proponents admit that the nuclear force seems to be incompatible with QCD[30]. A more detailed discussion of Wilczek's article appears later in this book in the chapter "And Yet, Why Do Scientists Believe in QCD?"

In fact, even today, the question of how the strong interaction explains the strong nuclear force is listed as one of the important unsolved problems in physics.[31,32]

The volume of external electrons

The volume[33] of the external electrons of a molecule inside a droplet is larger than their volume in a free molecule[34]. This property is a consequence of the screening effect that allows electrons to penetrate into neighboring molecules. The electrons of liquid droplet molecules thus partially overlap each other.

---

[30] Frank Wilczek, *Hard-core revelations*, NATURE, Vol. **445** 156 (2007). "Ironically, from the perspective of QCD, the foundations of nuclear physics appear distinctly unsound".

[31] Wikipedia list of unsolved problems in physics (October 2010.)

[32] In Physical Review Letters, 2007, Ishii, Aoki and Hatsuda published an article that shows preliminary results in which lattice QCD explains the nuclear force. However, their calculations are very far from being considered as the final word. For example they use a pion's mass of 0.53 GeV whereas the true value is about 0.14 GeV. Other unphysical mass values are also used in their article.

[33] A simplified definition of the electron volume is the volume of the orbit of the electron around the molecule or atom.

[34] J. B. Pendry, *The electronic structure of liquids*, J. Phys. *C*, **13**, 3357 (1980)

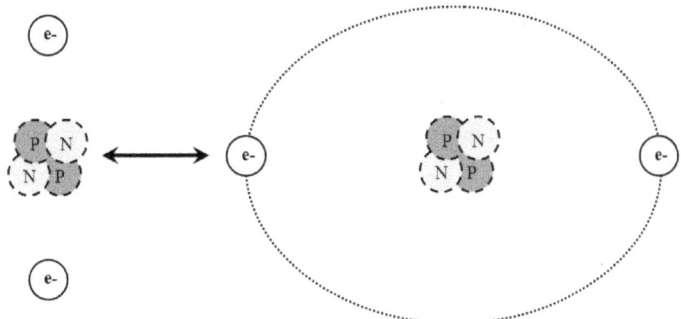

Figure 9. The larger volume of the electrons in liquid helium, caused by the attraction of electrons to the nuclei of neighboring atoms

A similar phenomenon was found in nucleons. In 1983, experiments discovered that the volume of nucleon quarks is larger for nucleons of a heavier nucleus[35]. This effect is called "the first EMC[36] effect" and it totally contradicted theoretical predictions published earlier.

The first EMC effect has been bewildering physicists up to the present day, since there is no consensus explanation accepted by the physics' community.[37]

Cross section curve

When a quite low energy beam hits an atom, it interacts with the electrons of the external shell. Then, according to quantum mechanical laws, as the beam's energy increases, the number of

---

[35] J.J. Aubert *et al.*, Phys. Lett. **123B**, 275 (1983). The graph on p. 277 shows that the *x* dependence of the structure function of iron is narrower than that of the deuteron. It follows that iron's quarks are enclosed in a larger spatial volume than that of the deuteron.

[36] European Muon Collaboration

[37] J. Arrington *et al.*, *New Measurements of the EMC Effect in Few-Body Nuclei*, J. Phys. Conference Series **69**, 012024 (2007). *"So while the experimental signature is clear, the interpretation of this effect is, at present, ambiguous."*

interaction events decreases. In such a case we say that "the cross section curve decreases."

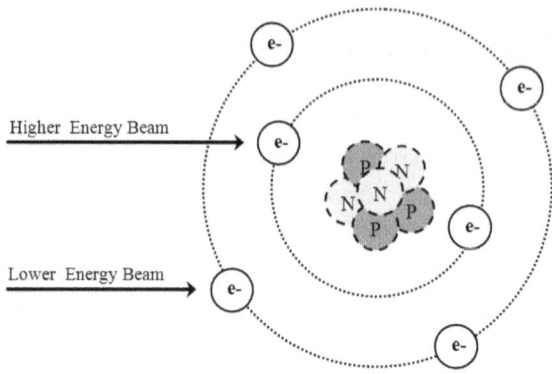

Figure 10. Interaction of a beam of electrons with the electrons in the atomic shells.

But when the energy of the beam is strong enough, then particles of the beam are able to excite the electrons belonging to the inner shells. Therefore, new participants enter the game, more hits are observed, and the cross section graph changes direction and begins to increase.[38]

This rise of the cross section curve as the beam energy increases and inner shells enter the process can also be observed in the shells of nucleons inside the atomic nucleus. It is well known that nucleons occupy shells inside the nucleus as electrons occupy shells in the atom. And indeed, when an electron beam hits a nucleus, a quite low energy electron beam can excite the nucleons in the external shell of

[38] C. J. Powell, *Cross sections for ionization of inner-shell electrons by electrons*, Rev. Mod. Phys., **48**, 33 (1976).

the nucleus, and when the beam energy is sufficiently high it begins to excite inner shells as well.[39]

A similar cross section curve is observed inside the proton. In experiments conducted before 1997 the cross section graph of electron beams hitting the quarks within the protons went down as energy increased. But a 1994-1997 experiment at the DESY labs in Germany, with the highest electron-proton collision energy to date, the measured cross section was higher than expected. These finding were followed by the publication of papers expressing astonishment.[40,41] Scientists decided to conduct additional experiments in order to assure that these results were not just statistical fluctuations.

Experiments conducted during the early 2000s in the Tevatron collider at Fermi National Accelerator Laboratory, Illinois, reached sufficiently high beam energies and confirmed that the cross section graph of proton-proton (and proton-antiproton[42]) collisions stops decreasing and begins to increase.[43]

This phenomenon remains unexplained.[44]

---

[39] I. Sick *et al.*, *Charge density of ^{40}Ca*, Physics Letters B, Volume **88**, Issues 3-4, (1979). p.245-248

[40] C. Adloff *et al.*, *Observation of Events at Very High Q2 in ep Collisions at. HERA*, Z. Phys **C74**, 191 (1997).

[41] J. Breitweg *et al.*, *Comparison of ZEUS Data with Standard Model Predictions for ep -> eX Scattering at High x and Q2*, Z. Phys **C74**, 207 (1997).

[42] An explanation about particles and anti-particles will be provided in the chapter "Particle Classification".

[43] K. Nakamura *et al.* (Particle Data Group), Journal of Physics **G37**, 075021 (2010). pdg.lbl.gov/2010/reviews/rpp2010-rev-cross-section-plots.pdf (p. 12)

[44] An attempt to explain this phenomenon was done by A. A. Arkhipov,

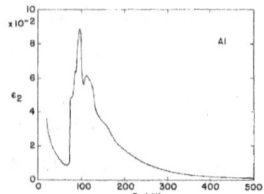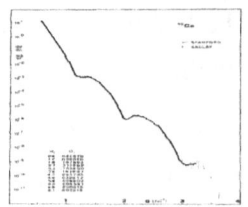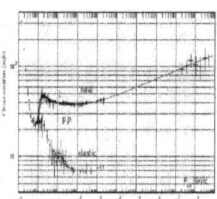

Figure 11. Regions of cross section curve rise in atoms (left), atomic nucleus (middle) and protons (right).

Pairs of particle and antiparticle

In 1947, Lamb and Retherford conducted an experiment on hydrogen atoms and found a small shift of energies compared to predictions based on pure quantum mechanics, now known as "Lamb shift." This result is well understood by a higher rank theory called quantum field theory. A simplified explanation of this shift is that the electronic state in the atom contains an additional pair of electron and anti-electron (called positron).[45]

The positron of the additional pair obeys the laws of electrodynamics and is evidently pushed from the atomic inner regions, because in these regions the nuclear field is only partially screened by electrons. Therefore, the positron, having a positive charge just like the atomic nucleus, should be located mostly in the atom's periphery.

---

arxiv.org/PS_cache/hep-ph/pdf/9911/9911533v2.pdf. This author uses a kind of force which is inconsistent with the Standard Model.

[45] For every massive particle there is an "anti-particle" which has the same properties as the massive particle but opposite charge. The positron is the anti particle of the electron.

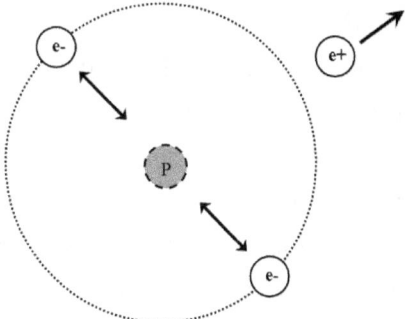

Figure 12. A state of the hydrogen atom containing an additional electron-positron pair. The positron is repelled by the atomic nucleus because both are positively charged

A similar phenomenon was found inside the proton. The proton has additional pairs of quark and antiquark beside the three valence quarks. Here antiquarks are measured explicitly and they tend to be in the proton's peripheral region.[46]

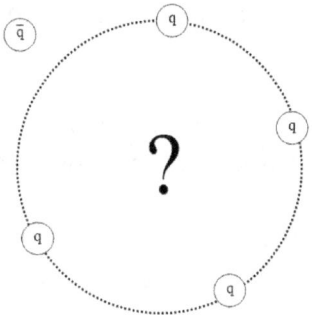

Figure 13. The antiquark is located in the peripheral region of the proton. This phenomenon is unexplained

---

[46] D. H. Perkins, *Introduction to High Energy Physics*, (Addison-Wesley, Menlo Park, CA, 1987). p.281. The smaller x-width of the antiquarks graph and the uncertainty principle prove that the antiquarks' volume is larger than that of quarks.

Proton form factor

In atoms there is probability to find the electron at a certain distance from the atomic center. It turns out that for the simple case of the ground state of the one electron hydrogen atom, the probability to find the electron close to the center is higher than the probability to find it at the periphery. This phenomenon is explained by quantum mechanics using the fact that the electromagnetic force that attracts the electron to the positively charged nucleus *decreases* while the distance between the electron and the nucleus *increases*. This issue is agreed by the entire physical community.

What is the probability to find the quarks at a certain distance from the nucleon center?

In the proton's case information on the spatial quark distribution is based on measurements. The analysis of the experimental data yields a mathematical quantity called "form factor." The form factor can be used for deriving the probability to find the quarks at different points inside the proton.

It was found that in the proton, similarly to the hydrogen atom, the probability to find the quarks near the proton center is higher.[47]

The proton spin crisis

It is well known that the electrons in each atomic state exist in several configurations.[48] We will discuss this issue in detail later in this book. During the 1940s Wigner and Racah developed useful mathematical tools for calculating these configurations. As a result

[47] D. H. Perkins, *Introduction to High Energy Physics* (Addison-Wesley, Menlo Park, CA, 1987). p. 194-196

[48] A.R.P. Rau, *Astronomy-inspired Atomic and Molecular Physics* (Kluwer Academic, Dordrecht, 2002). p. 16

of these configurations, the sum of the electron spins[49] is much smaller than the total spin of a single configuration.[50]

A similar phenomenon occurs in the proton. In 1987 EMC measured the sum of the spins of quarks inside the proton, and found that the total was much smaller than the overall proton spin. This phenomenon is called the proton spin crisis and is considered one of the most important unsolved problems in physics.[51]

Baryon number conservation law

It is well known that the number of atoms is conserved in chemical interactions. This is because chemical processes involve the external electron shell of each atom, while the atomic nucleus remains unchanged. Therefore, the number of atoms does not change in chemical processes.

A similar phenomenon is found in baryons. Baryons are particles belonging to the proton and the neutron family. Different baryons have different setup of the three valence quarks. In all experiments to

---

[49] Spin is a fundamental characteristic property of particles. It is a kind of a quantum mechanical rotating top. The spin of a system of particles is the sum of spins of the individual particles, where the summation obeys quantum mechanical laws.

[50] The reason is that multiple configurations lead to cancellation of a major part of the contribution of the electrons' spin to the total angular momentum of the atom, because the spin is coupled to the spatial angular momentum and in different terms it takes opposite directions.

[51] Wikipedia list of unsolved problems in physics. en.wikipedia.org/wiki/List_of_unsolved_problems_in_physics

date, the number of baryons is conserved. This is inconsistent with theories that predict proton decay.[52]

Tensor force between nucleons

According to the electromagnetic equations, when an electric charge moves in a loop, or when the charge has non-zero spin, it creates an axial magnetic dipole. The axial magnetic dipole is an axial magnet with two poles. When two axial magnetic dipoles are positioned near one another, they apply a force in a particular direction, which depends on the relative orientation of the axial magnetic dipoles and on the distance between them. This force is called "tensor force." In most cases, the direction of the tensor force is not parallel to the line connecting the dipoles.

Some atoms, like the hydrogen atom, have non-zero spin and their electrons create axial magnetic dipoles. When two hydrogen atoms are positioned relatively far from one another,[53] then *in principle* a small tensor force is exerted by one atom on the other.

---

[52] Howard Georgi and Sheldon Glashow, *Unity of All Elementary-Particle Forces*, Physical Review Letters, **32** 438 (1974).

[53] When two hydrogen atoms come close to each other they create hydrogen molecule (H_2) and destroy the magnetic axial dipoles of the combined system.

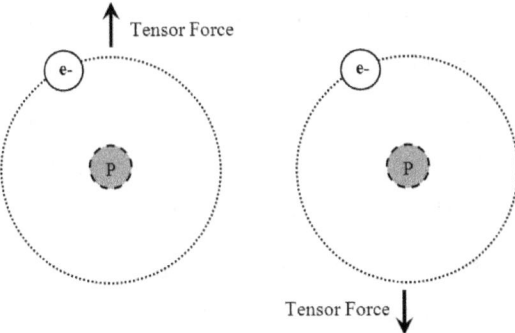

Figure 14. Atoms with spins generate magnetic dipoles and apply tensor forces on one another

The tensor force appears more effectively with nucleons. Protons and neutrons have non-zero spin, and they apply a tensor force on one another. As far as we know, the question as to what causes the tensor force is still open.

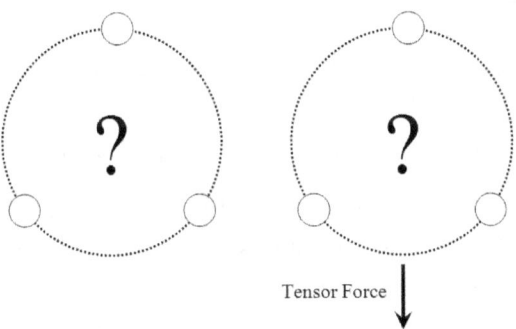

Figure 15. Nucleons apply tensor force one on another.

Strong CP problem

Electromagnetic force, operating between electrons and protons, obeys the law of C (charge conjugation) and P (Parity) conservation. This means that in a world in which matter becomes anti-matter, or in a mirror world, where right and left are interchanged, the same electromagnetic laws apply.

The same applies to strong interaction: experimental findings indicate that the strong interaction is invariant under charge conjugation and under parity transformations.

However, the current theory of particle physics, the standard model, predicts that strong interaction should violate CP. In the literature, this problem is called the "strong CP problem," and is considered one of the most important unsolved problems in physics.[54]

Interaction with light

According to electromagnetic theory, when a beam of light hits an atom, it interacts only with electric charges. And indeed, light interacts with the electrons of the atom, and energy may be transferred from the photon to the electron. This effect is called the "photoelectric effect".

About 50 years ago, physicists discovered that energetic photons interact with the quarks more intensely than expected by the quarks' electric charge. Moreover, the interaction of energetic photons with both protons and neutrons is very similar, although the protons are electrically charged, and the neutrons are not. Physicists agree that this cannot be explained by the quarks' electric charge. A theory explaining this phenomenon was published in the 1960s,[55] but was later removed from most of the textbooks and is not considered as a

---

[54] Wikipedia list of unsolved problems in physics. en.wikipedia.org/wiki/List_of_unsolved_problems_in_physics (November 2010)

[55] T. H. Bauer, R. D. Spital, D. R. Yennie and F. M. Pipkin, *The hadronic properties of the photon in high-energy interactions*, Rev. Mod. Phys., **50**, 261-436 (1978).

valid theory. As of today, there is no other accepted explanation of this fundamental phenomenon.

Let's compare

Table 1. A comparison between electromagnetic force and strong interaction.

	Electromagnetic force and electrons	Strong interaction and quarks
1	Holds electrons inside an atom or a molecule by means of a relatively strong force.	Holds quarks inside a nucleon by means of a strong force.
2	Holds molecules within a liquid droplet by means of a much weaker force (the van der Waals force).	Holds nucleons inside a nucleus by means of a much weaker force (the strong nuclear force).
3	Cancels rapidly when the molecules are far apart from each other.	Cancels rapidly when the nucleons are far apart from each other.
4	Liquid molecules have a quasi-constant density.	Nucleons within atomic nucleus have a quasi-constant density.*
5	The volume of electrons of a molecule inside a liquid droplet is larger than that of a free molecule.	The volume of nucleonic quarks inside a heavy atomic nucleus is larger than that of the deuteron (first EMC effect).*

6	The graph describing distance dependence of the molecular force potential looks like a Ski Jump curve.	The graph describing distance dependence of the nuclear force potential looks like a Ski Jump curve.*
7	If the collision energy is high enough, the cross section graph increases.	If the collision energy is high enough, the cross section graph increases.*
8	Due to quantum field theory, atoms contain a very small amount of additional pairs of electron and positron.	Nucleons contain additional pairs of quark and antiquark.
9	According to electrodynamics the positron of this pair tends to be located away from the center of the atom.	Experiments show that antiquarks tend to be located away from the center of nucleon.*
10	The electrons tend to be close to the atomic center.	The quarks tend to be close to the proton center.*
11	Electron spins do not accumulate.	Quarks spins do not accumulate (proton spin crisis).*
12	The number of atoms doesn't change in chemical interaction.	The number of baryons doesn't change in any interaction.*

13	In principle, two spatially distant atoms with non-zero spin apply a tensor force on one another.	Nucleons have non-zero spin and two nucleons apply a tensor force on one another.*
14	The electromagnetic force does not violate CP.	The strong interactions do not violate CP (strong CP problem).*
15	Photons interact with electrons.	Photons interact strongly with quarks.*

* Unexplained experimental result

It turns out that the characteristics of molecules/atoms are amazingly similar to those of nucleons, if we substitute an atomic nucleus for a droplet of molecules, nucleons for molecules/atoms, and quarks for electrons. Most of these experimental results have remained unexplained, possibly because they are incompatible with QCD, the currently admitted theory.

Is nature trying to tell us something?

Can you, the reader, deduce the structure of the nucleons?

Can you deduce the kind of forces which hold quarks inside nucleons?

Chapter 2: Non-Conservation of Knowledge

One of the consequences of the outstanding development of scientific knowledge in modern times, and its extensive ramifications, is that no scientist could possibly be familiar with all the scientific facts, discoveries and information, not even within his or her own field of specialization.

The strength of a scientific community lies in its capacity to preserve acquired, validated knowledge and develop it further. It is therefore quite hard to understand how, when physicists discover the extent to which QCD fails to explain a large number of experimental findings, their common reaction is surprise, which then turns into disbelief, suspicion and sometimes even aggressiveness.

Most physicists are well aware of the fact that QCD is unable to explain the strong nuclear force. The strong nuclear force and other phenomena, such as the "proton spin crisis" and the "strong CP problem," are famous "waiting to be solved" problems.

Many other problems – some of which will be presented in this book – actually contradict QCD, but for some reason, have never been properly addressed.

Interaction of photons with nucleons

The interaction of electromagnetic waves with electric charges has been considered as a milestone in physics since the 19th century with the introduction of Maxwell's equations. Countless devices we use in our daily life are based on this phenomenon. The development of quantum mechanics at the beginning of the previous century and the discovery of the photoelectric effect extended this knowledge to the photon.[56]

Some fifty years ago, scientists were amazed to discover that energetic photons were reacting to the forces inside the nucleons as well.[57,58]

From a theoretical perspective, the significance of this phenomenon is just as important as the interaction of photons with electric charges. In fact, the interaction of photons with strongly interacting particles (called quarks) seems to provide an important clue for the existence of a common denominator between electromagnetic and strong interactions. The above mentioned long list of similarities between electromagnetic systems of electrons and strongly interacting systems of quarks, "cries out" for finding such a common denominator.

---

[56] In 1905, Albert Einstein explained the photoelectric effect by describing light as composed of discrete quanta, now called photons.

[57] T. H. Bauer, R. D. Spital, D. R. Yennie and F. M. Pipkin, *The hadronic properties of the photon in high-energy interactions,* Rev. Mod. Phys., **50**, 261-436 (1978).

[58] This photon related force is much stronger than its well known interaction with the charge constituents of the nucleons. Moreover, although the proton and the neutron have different electric charge, they interact about the same with energetic photons.

The newly discovered phenomenon was very fashionable during the 1960s, starring in graduate physics study programs and attracting scientists' comprehensive efforts to explain it. But as the years passed, and no theory seemed to be capable of providing a satisfactory explanation, the phenomenon fell into oblivion. A review of four[59,60,61,62] of the particle physics textbooks currently used in graduate schools, shows that this issue is not even mentioned.

Furthermore, in Wikipedia, the most comprehensive encyclopedia existing today, this phenomenon doesn't appear in the Photon topic. In fact, it appears in Wikipedia only once, in a short, vague comment.[63]

It turns out that such a fundamental phenomenon, of the same significance as the photoelectric effect itself, receives only negligible

---

[59] D.H. Perkins, *Introduction to high energy physics*, (4th ed. Cambridge University Press, 2000).

[60] D.J. Griffiths, *Introduction to elementary particles*, (2nd, rev. ed. Weinheim : Wiley-VCH, 2008)

[61] F. Halzen and A.D. Martin *Quarks and leptons*, (New York : Wiley, 1984)

[62] Fayyazuddin and Riazuddin, *A modern introduction to particle physics*, (2nd ed. Singapore : World Scientific, 2000)

[63] en.wikipedia.org/wiki/Vector_meson_dominance. *In physics, vector meson dominance was a model developed by J. J. Sakurai in the 1960s before the advent of QCD in order to describe interactions between photons and hadronic matter. In particular the hadronic components of the photon polarization tensor consist of the lightest vector mesons ρ, ω and φ. Therefore interactions between photons and hadronic matter occur by the exchange of a hadron between the dressed photon and the hadronic target.*

attention. It is nothing less than a disgrace that most physicists are not even aware of this property of the photon.

The first EMC effect

A paper published in January 1983[64] reported experimental findings according to which the volume of nucleon's quarks in heavy nuclei is larger than their volume inside the deuteron. The authors indicated that this phenomenon contradicted all theoretical predictions.

Since then, no scientific theory explaining this result has been introduced.[65] This intriguing phenomenon is not even mentioned in graduate study programs and most physicists would hardly be capable of understanding its nature even after searching the literature.

As far as we know, the only reference to the phenomenon in Wikipedia is one vague comment.[66]

The proton-proton cross-section curve

Two papers published in 1997[67,68] reported an unexplained increase in the number of events for very high energy electrons/positrons scattering on protons, compared to the expected value. Nearly every physicist would agree that the only plausible explanation for this

---

[64] J.J. Aubert *et al.*, Phys. Lett. **123B**, 275 (1983).

[65] J. Arrington *et al.*, J. Phys. Conference Series, **69**, 012024 (2007).

[66] en.wikipedia.org/wiki/European_Muon_Collaboration. *In 1983, EMC discovered that nucleons inside a nucleus have a different distribution of momentum among their component quarks. This is the original so-called "EMC Effect".*

[67] C. Adloff *et al.*, *Observation of Events at Very High Q2 in ep Collisions at. HERA*, Z. Phys **C74**, 191 (1997)

[68] J. Breitweg *et al.*, *Comparison of ZEUS Data with Standard Model Predictions for ep -> eX Scattering at High x and Q2*, Z. Phys **C74**, 207 (1997)

finding is the existence of additional massive objects inside the proton, which contradicts the consensus theory.[69] This experimental data was regarded as inconclusive because of the small number of events.

An analogous experimental finding has been verified beyond any doubt in an experiment at the Tevatron particle collider in the beginning of the 2000s. This machine measures proton-proton and proton-antiproton collisions at very high energy. The results clearly show that for high energy both the elastic and the total cross-section curves stop decreasing and begin to increase.[70]

There are hardly any papers dealing with this fundamental issue, although the consequences of this result would totally shake the common belief about the structure of nucleons. A search in Wikipedia shows no mention of this finding, either.

Claims stating that the increase of the proton-proton cross section curve is inconsistent with QCD have already been published a decade ago[71]. No adequate discussion of this problem can be found in the scientific literature.

It turns out that a perfectly valid experimental finding, which seems to contradict the dominant theory, is universally ignored.

---

[69] *"If the results are not a statistical fluke, new physics has been observed. One possibility is that our understanding of what's inside the proton is somehow wrong".* Frank Sciulli, Columbia University News, 1997.

[70] C. Amsler *et al.* (Particle Data Group) Physics Letters **B667**, 1 (2008). p.12. pdg.lbl.gov/2009/reviews/rpp2009-rev-cross-section-plots.pdf

[71] A. A. Arkhipov, arxiv.org/PS_cache/hep-ph/pdf/9911/9911533v2.pdf.

The NIH syndrome

The NIH (Not Invented Here) syndrome is a term used to describe a culture that avoids using knowledge because of its external origins. This term was coined to describe the attitude of large high tech companies in the past, but it can unfortunately characterize many particle physicists who ignore physical theories attributed to neighboring research fields. Established physical evidence other than that conceived by particle physicists is often totally ignored, and sometimes even overtly scorned.

For example, historically, QCD color force was invented in order to solve the puzzle of the Δ^{++}, Δ^- and Ω^- particles. During the 1960s, particle physicists thought that these particle properties contradicted the Pauli exclusion principle, which is considered to be a fundamental concept in quantum mechanics.[72] A simplified formulation of Pauli's principle states that two identical fermions[73] cannot exist in the same quantum state. In Δ^{++}, Δ^- and Ω^- there are three identical fermions in the external shell (three u quarks in the case of Δ^{++}, three d quarks in Δ^- and three s quarks in Ω^-). The total spin and parity of each of these particles is $3/2^{+}$[74] and the total isospin of the Δ baryons is 3/2. Particle physicists considered this relatively stable combination impossible without the invention of QCD colors.

However, such combinations do exist in nature, and this is well known to nuclear physicists.

---

[72] F. Halzen and A. D. Martin, *Quarks and Leptons* (Wiley, New York, 1984). p.5

[73] Electrons, nucleons and quarks are examples of fermions

[74] Parity is an important property of a quantum state. Its standard notation is a superscript + (-) for a positive (negative) parity. See the "Terminology" appendix.

35

It is well known that the atomic nucleus contains nucleons, and they reside in shells.[75] Let's examine several nuclei that have 31 nucleons: silicon ^{31}Si, phosphorus ^{31}P, sulfur ^{31}S, and chlorine ^{31}Cl. All of them have 14 protons and 14 neutrons in their internal closed shells, and 3 nucleons in their nucleus' external shells. Silicon has three neutrons, phosphorus has two neutrons and one proton, sulfur has two protons and one neutron and chlorine has three protons in its nucleus external shells.

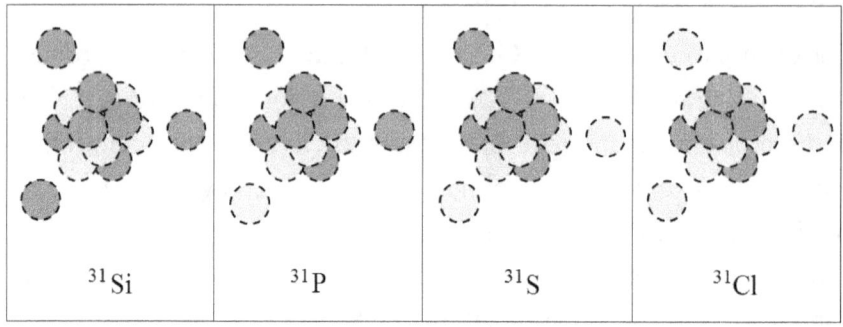

^{31}Si ^{31}P ^{31}S ^{31}Cl

Figure 16. The structure of the nuclei of silicon, phosphorus, sulfur and chlorine that contain 31 nucleons

Quantum states of these isotopes demonstrate an analogy to the baryons Δ^- (with ddd quarks), neutron and Δ^0 (udd), proton and Δ^+ (uud) and Δ^{++} (uuu). Here too, nucleons are fermions that obey the Pauli exclusion principle and a spin, parity and isospin correspondence exists between the nuclear and the quark states.

If we follow the "proof" that particle physicists provided that Δ^{++} and Ω^- cannot exist without colors then we will see that the same "proof" holds for the ground state of the nuclei ^{31}Si and ^{31}Cl.

---

[75] M.G. Mayer, *On closed shells in nuclei*, Phys. Rev. **74** (1948). p.235-239

Why then is there a rather stable ^{31}Si isotope with three neutrons in its external shell? And why does the isotope ^{31}Cl with three such protons exist?[76] Both isotopes have spin-parity $3/2^+$ and isospin $3/2$ like Δ^- and Δ^{++}. If such particles exist in nature without colors, then why is the invention of colors needed to explain Δ^{++}, Δ^- or Ω^-?[77]

The disregarded theory

It is well known that electrons in the atom are structured in shells. The first shell is called "1S" and it may contain two electrons at most. Too many textbooks claim that in the helium atom "ground state"[78] the two electrons reside only in the "1s" shell.[79,80] This is taught as a scientific truth and provided in countless academic web sites.[81,82,83]

This is simply wrong. It was very well known that the electron shells in the atom are far more complicated than is generally admitted. Calculations based on a well-established mathematical theory developed by Wigner and Racah during the 1940s have proven this.

---

[76] These isotopes disintegrate because of weak interactions.

[77] E. Comay, *On the Quantum Mechanical State of the Δ^{++} baryon*, Prog. in Phys. **1**, 75 (2011). tau.ac.il/~elicomay/dpp_fin.pdf

[78] Ground state of an atom is its lowest energy state

[79] John Olmsted, Gregory M. Williams, *Chemistry, the molecular science*, (John Wiley & Sons, Inc), p. 310

[80] Frank H. Shu, *The physical universe: an introduction to astronomy*, p.51

[81] hyperphysics.phy-astr.gsu.edu/hbase/quantum/helium.html

[82] wiki.brown.edu/confluence/download/attachments/29133/Helium+and+Calcium.pdf

[83] quantummechanics.ucsd.edu/ph130a/130_notes/node35.html

Physicists knew 60 years ago that electrons are stored in several configurations simultaneously.[84]

Physicists may argue that the claim that the electrons are in one shell is a legitimate approximation of reality. However, this is not so. First of all, the helium configurations were calculated and it appears that even in this simple two-electron atom, there is quite a high probability that the electrons are stored in several configurations.[85]

But more important, for our discussion, is that physicists use the same wrong arguments to explain the physical properties of quarks. Furthermore, it seems that they do not take into account the profound reasons that multiple configurations reduce energy level. As we will see in this book, if they would adopt the well-established configuration concept, they would be amazed to discover that it offers a solution to one of the major problems of the last few decades: the proton spin crisis. Furthermore, it provides better understanding of the structure of baryons and the properties of Δ^{++} and Ω^- baryons.

Sixty years ago, after physicists found that nucleons occupy shells in the nucleus, Racah extended the applications of his mathematical theory in order to calculate the states of nucleons in the atomic nucleus.[86] It is reasonable to believe that Racah would have tried to extend his theory further in order to explain the quark properties after they were discovered during the 1960s. Unfortunately, Racah

---

[84] G.R. Taylor and R.G. Parr, *Superposition of configurations: The helium atom,* Proc., Natl.Acad. Sci. USA **38**, (1952). p.154-160

[85] AW Weiss, *Configuration Interaction in Simple Atomic Systems*, Phys. Rev. **122**, (1961). p.1826–1836

[86] G. Racah, *Directional correlations of successive nuclear radiations*, Phys. Rev. **84** (1951). p.910-912

passed away in 1965 and no one made any further efforts in this direction.

Today, publishing articles in this field in mainstream physics journals is nearly impossible. In fact, as an important scientific editor candidly said, such articles could not be published because they could not go through the mandatory procedure of peer review; no scientist he knew of would be capable of reviewing such papers.

The search for new particles

Four years after Dirac published a paper that was used for predicting the existence of the positron, the particle was discovered. The non-discovery of the positron would have most likely resulted in the collapse of Dirac's theory, which predicted the existence of anti-matter.

In 1961 Gell-Mann and Ne'eman used the mathematical model of the SU(3) symmetry group to predict the existence of the Ω^- particle. The particle was discovered three years later. The non-discovery of this particle would have disqualified Gell-Mann and Ne'eman's entire theory, which constitutes the foundation for the quarks.

On the other hand, QCD predicts the existence of many types of particles and matter, such as dibaryons, glueballs, pentaquarks and strange quark matter. Several decades have elapsed since the theoretical predictions of these objects and yet, despite huge investments of experimental efforts, the existence of none of these entities has ever been confirmed.

But for some reason, although these particles have not been discovered, the theory used for predicting them has not been questioned. Within a few decades, after the experimental search stops, the unsuccessful predictions will probably be forgotten.

Chapter 3: The Photon also Blinks

When we finished writing the book in front of you, I felt that the human and sociological side was missing, and in order to fill that gap, this parable was written. Some readers asked me why I exaggerated the scientists' responses, and presented them in a ridiculous light. Well, everything written here did occur in the appropriate parable, every sentence told here was said or written in e-mails without any attempt to exaggerate.

The story occurred between October 16, 2009 and the beginning of 2011.

We stood there on the hill, my father and I. Below us was the Nucleon Lake. We were watching hundreds of scientists wandering the area.

The high concentration of scientists at Lake Nucleon was extraordinary. "What is special about this lake?" I asked.

"The scientists believe that the lake was formed by extraterrestrial beings." My father said. "They believe that there is no other explanation for the lake's presence. They also claim that there are fish species in the lake that are genetically unlike any other on earth. That's why they think that the lake was formed by water and creatures brought from beyond earth."

"Wow", I said with astonishment, looking at the scientists' dirty work. Some of them entered the lake's black water with special diving goggles on their heads.

"What are they doing in there?" I asked.

"Exploring the lake is very difficult," my father said. "When entering the lake, almost nothing is visible, and the exploration is done very slowly. This lake has been explored for forty years, and the scientists still have no explanation as to how it was formed and what is in it. I too have dedicated years to exploring it."

"And what have you found?" I asked.

He pointed at a distant mountain peak.

"Almost thirty years ago, in 1983," said my father, "I found the way to the peak of that mountain. From there I could see, to my amazement, that the lake is connected to the ocean by a number of streams. The water source is the ocean itself, and the fish in the lake come from the ocean."

I looked at him, disbelieving. "So why haven't you told them?!" I asked, and pointed at the group of scientists that continued to wander around below us.

"I tried," answered my father. "They just don't want to listen. At the time, I managed to convince some of them to climb with me to the mountain peak, but they quit on the way. It took me time to understand that they lacked climbing skills and they didn't want to admit it." He stalled, and added: "I guess they spend most of their study years learning diving in the lake, and they completely neglected mountain climbing."

There was a very large gathering of scientists below by the lake. "Why are there so many scientists here?" I asked.

"The scientists have a problem with not knowing what the fish feed off in this lake. They invented a theory that there is a small fish that all the other fish feed off. This fish is called the Higgs fish. The fish hasn't been found yet and there is a special effort now that involves scientists coming from around the world to find it in the lake."

"This small fish is so important?" I asked.

"They think so. Most of them think that if it isn't found by 2014 then it doesn't exist and their entire theory falls apart. I'm sure there is no such fish, as according to their theory, the Higgs fish has no gills, and I proved that there are no fish without gills."

After speculation, I asked: "And once they don't find the Higgs, will they agree to join you on that mountain?"

"I hope so, but I expect not. I expect that by then a new theory will pop up and explain why that fish is unnecessary, or they'll find a fish with gills and decided that that is the Higgs."

I examined the mountain. It was very high. "In order to see those streams, does one have to go to the top?" I asked. My father nodded. He has been climbing mountains for decades, whereas I have never officially learned to climb. But I am a good sportsman, and have won many competitions, included world championships. I decided to accept the challenge.

"Can you guide me there?" I asked.

The way was difficult. He skipped ahead with joy that I haven't seen in him in years, guiding me in how to advance. After a number of months, during which I advanced only a little, the mountain peak still looked very distant. I stopped, looked around, and saw a hill, not too far away, that seemed a good observation point.

"Maybe we'll reach there?" I asked him, and pointed to the hill. "It will take me a long time to reach the actual peak. Maybe something is visible from that hill."

The view we had from that observation point was amazing. Some of the streams that my father saw almost thirty years ago were visible. It was clearly possible to see how the water exited the ocean and flowed into the lake.

"The course that the streams take is visible from higher up, as is how the fish enter the lake from the ocean," said my father, "but even from here much is visible." He added in astonishment.

We returned back down, and I decided to try and meet the scientists exploring the lake.

This turned out to be a complex task. I suppose the diving goggles that they wore when entering the lake damage their hearing and eyesight, otherwise it isn't clear why they ignored me. Eventually I managed to get an appointment with one of the scientists for a few hours. I was lucky that he was a senior, good-hearted scientist who was very knowledgeable, and happy to share his knowledge.

"Can you tell me why you believe that the lake was formed by an extraterrestrial agent?" I asked.

He was very happy to answer. "Both the water and the fish found here cannot be created on earth, therefore we have no doubt that they were brought here by an external source. There is a wide variety of proofs. One of the things we discovered was that all the fish here have three eyes."

That didn't amaze me particularly. My father had already explained to me why the fish have to be three-eyed in order to survive in the lake.

"The ocean also has fish with three eyes." I pointed out.

"True," said the scientist, "but fifty years ago we discovered a fish here that we called Ω-, that has three eyes all the same color, and that all look in the same direction. Fish like these can't be created on earth."

I had read up on the subject on the internet, and this claim didn't surprise me.

"But the ocean has fish with three eyes, all the same color, all looking in the same direction." I said.

He frowned. "There aren't any," he said.

"There are", I said, "I can show you two like this on Wikipedia."

He knew my father, and spoke slowly and deliberately. "All the scientists in the field, except your father, think that there can't be fish like these in the ocean."

I apologized. He continued.

"There are other kinds of fish that don't exist anywhere else." said the scientist. "We discovered that sometimes, when we strike the lake with force, two regular fish jump out, and one special fish called Gluon. There is no Gluon fish anywhere on earth."

"And how do you know that the fish is a Gluon?" I asked, "Did you ask it?"

He laughed. "That's a good question", he said, "You could have been a wonderful scientist. I recognize wasted talent. Well, it turns out that the fish has one fin."

"The Photon fish also has only one fin." I said.

"True, but this fish also blinks!" said the scientist victoriously.

"The Photon also blinks." I said to him.

"Mistake", said the scientist, "the Photon only moves its tail right and left. It doesn't blink."

"It does blink if it's struck hard enough." I said.

The blinking property of the Photon was indeed discovered in the 1950s, and surprised the scientists that were sure that the Photon could only move its tail right and left. At the time an attempt was made to explain the Photon's blink, but the explanation was considered dubious even by the scientists. Today the blinking phenomenon and its explanation don't appear in the literature, despite the Photon being the most common fish in the ocean.

The senior scientist remembered the phenomenon after I told him the story.

"OK", he said angrily, "It's not a Photon, and that's proven!"

I stopped speaking and allowed him to finish his lecture without disturbances.

In the following days I discovered that there is a very short way to another, even better than the previous, observation point. It was one half-hour away at the most, perhaps forty five minutes, for scientists experienced in mountain climbing. I made another appointment with the senior scientist. I told him that I want to show the discovery of my father.

"My father discovered that the lakes water comes from the ocean." I told him. "Come with me for a short walk, and I'll show you." He joined me and explained to me as we advanced towards the observation point why it isn't possible for the water to come from the ocean.

When we were nearing the hill, some of the streams flowing towards the lake were already visible. He stopped, and looked in wonderment at the streams. He continued a little and saw another span of them.

"Another five minutes of walking and you'll be able to see the streams and the ocean clearly." I said.

And then he stopped climbing. "Sorry, but I've just remembered that I have to return." He started on his way back down quickly. As he walked he called back out something like, "If you mention my name in connection to this, then state that I don't agree with anything your father says!"

I told this to my father. He wasn't surprised. "For almost thirty years I've been trying to bring them there." He told me. "They prefer not to know."

There are many fish in the lake that behave exactly like the fish in the ocean. But scientists are sure that the fish came from another planet, and that they have to behave in the opposite manner from the fish from the ocean. That's why they were astonished each time that the fish in the lake behaved naturally like the fish in the ocean, and labeled this behavior as "an unexplained mystery."

I continued trying to interest them in the subject. I asked one of them why he wasn't interested in accompanying me to the hill. "The trip there takes just one half-hour, and from there you could understand the explanation for these mysteries immediately." I told him.

"The fact that these are mysteries doesn't pose a problem," said the scientist. "It is known that the water here is black, and that almost nothing can be seen in it, and the research is very slow. Therefore, many mysteries are still unexplained, and we assume that many of them will remain unexplained forever."

This answer was a surprise for me. I asked two other scientists who gave a similar reply, one of them added: "If you would have said that you can see from there that the water comes from another planet – I would have joined."

I asked my father if he was prepared to say that it's possible that the water comes from another planet, so they would come to see it from the top of the hill. He laughed and said that this was suggested to him in the past.

"That's swindling, and I'm not a swindler." he said.

While I was travelling in the mountains I met a famous scientist who was wandering there. He wasn't an expert on Nucleon Lake, but he had great knowledge of the mountains. When he heard my father's story from me, he was very surprised and said: "After all, your father is a bona fide mountain-climber." He said. "I don't understand why they don't come with him to the top of the mountain."

"Maybe you should tell them yourself?" I asked. But he refused.

Another famous scientist that I met there in the mountains was very nice, and even agreed to join me on the hill to see the phenomenon. He climbed up with me, looked down in amusement, and said: "What I see gives me great pleasure, but I'm not an expert on the lake itself, only of the streams and the ocean, so, it would be very irresponsible of me to give my opinion."

After deliberating on the subject for some time, I approached my father. "Dad, I have an idea. Let's open up a small coffee shop on the hill. Good sportsmen that aren't mountain climbers can get there within one to two hours. I'm sure that many would be happy up here, drinking beer, looking at the beauty of the streams, peaking at the scientists wandering around below."

My father agreed, and that's how this book was created. The next chapters will guide you to that hill, and after reading it, you too can peak at the view reflected from the same hill, and if you want to climb further up the mountain – you can read the entire book.

Chapter 4: The Scope of this Book

The 20th century witnessed the development of a branch of physics called quantum mechanics, resulting in major technological breakthroughs. In the 1950s and 1960s, physicists were aware of the existence of two particles considered impossible according to quantum mechanics. These particles were the Δ^{++} and the Ω^-.

During the 1960s it was already known that protons and many other particles are composed of smaller entities, the quarks. The particles Ω^- and Δ^{++} were known to be composed of quarks as well, but as discussed earlier in this book, the quark combinations and their properties did not seem coherent. At that time, it was not known that the quarks carry only half of the proton mass, and physicists thought that the quarks were the only *massive* objects inside the proton. Even today QCD supporters agree to the claim that quarks are the only kind of massive constituents of the proton.

The discovery of Ω^- and Δ^{++} seriously challenged the knowledge acquired up until then, seriously enough to motivate scientists to concentrate their efforts on the development of a new physical theory, based on a series of fantastic assumptions, describing forces and particles unlike anything known up to that point in time. The theory, called QCD, a central pillar of the standard model, won unshakable status as far back as the 1970s, in spite of a long series of incompatibilities with experimental findings (described in detail later on in this book).

Eliyahu Comay was a physics student during the 1960s at the Hebrew University of Jerusalem, where the eminent physicist Yoel Racah, was a revered figure. The physics program at the Hebrew University was particularly focused on the theory Racah developed in parallel with Wigner. Comay recognized, in the early 1970s, that Ω^- and Δ^{++} can be naturally explained by the theory developed by Wigner and Racah, in combination with the basic laws of quantum mechanics. Comay also realized that the masses of particles composed of quarks were consistent with the laws of quantum theory, established long before the invention of QCD.

Comay further specialized in nuclear physics. This was how he discovered that QCD's description of the forces inside the atomic nucleus was incorrect.

In 1983, Comay made an astonishing discovery, critical for understanding the physical properties of quarks, and the fundamental role of photons in strong interaction. This discovery allowed him to formalize the equations of quarks and to develop an alternative model, explaining familiar phenomena in an amazingly simple way, including phenomena which cannot be explained by QCD.

The scope of Comay's model
The experimental findings below are explained in Comay's model. These results do not have accepted explanation, and they seem to contradict QCD (all the terms below will be explained later in this book):

- Protons and neutrons behave similarly when a hard photon hits them.

- Protons and neutrons interact strongly when a hard photon hits them.

- The first EMC effect.

- The proton spin crisis.

- The strong CP problem.

- The potential vs. distance graph of van-der-Waals and strong nuclear forces are similar.

- Nucleons within the atomic nucleus have a practically uniform density.

- The nuclear tensor force and its sign.

- Antiquarks have a larger volume inside nucleons.

- The neutron's negative electric charge tends to be found in external regions.

- Slow decrease of the proton-proton cross section in an appropriate high energy region.

- Increase in very high energy proton-proton elastic cross section.

- Proton form factor.

- Pentaquarks were not found.

- Strange quark matter was not found.

- Glueballs were not found.

- Dibaryons were not found.

Comay's model explains other findings that do not contradict QCD, although some of them don't have an explanation yet:

51

- Linear momentum of quarks accounts for only half of proton momentum.

- Quark confinement.

- The properties of the Ω^- and the Δ^{++} baryons.

- The three jet event.

- Meson radius relations.

- The relation between the proton's radius and the pion's radius.

- Problems with mass differences between mesons and baryons.

- The strong force ceases to act at a certain distance (cutoff).

- π^0 decay.

- Mesons are not confined inside the nucleon.

- Baryon number conservation law.

- Proton decay was not observed.

- The quarks u,d are heavier in baryons rather than mesons, and quarks having other flavors are heavier in mesons rather than baryons.

- Genuine Yukawa particle and the Higgs particle were not discovered. In fact, Comay shows that there is no elementary point-like spinless particle.

- Dirac monopoles were not discovered.

Most of Comay's work was conducted and published during 1980s and 1990s. A concluding article that summarizes most of his findings regarding his model was published in 2004.[87]

Using everyday language, this book will cover many topics, some highly advanced, in quantum mechanics, quantum field theory, Wigner and Racah calculus and their tremendous impact on the strong force, and more. This will allow us to understand the solution to more than a dozen unsolved problems listed above, some of which feature in the list of the most important unsolved problems in physics.

Further, we will try to explain the historical events that brought the particle physics to its current bizarre situation.

Clarifications

Apart from special and general relativity, quantum mechanics is one of the most important development in 20th-century physics. During the 1940s and 1950s the development of quantum field theory allowed physicists to predict phenomena with outstanding accuracy, to a much larger extent than ever before. However, since the emergence of QCD in the 1970s, it would be fair to say that the branch of particle physics is in trouble.

The entire standard model is not challenged here. It would be reasonable to assume that many parts of the model, unrelated to QCD, could be valid, for example the notion of three generations of elementary particles or the Drell-Yan process, which Comay considers good to be and real achievements in physics. Comay

---

[87] *A Regular Monopole Theory and its Application to Strong Interactions*, Published in "Has the Last Word been Said on Classical Electrodynamics?" (Rinton Press, NJ, 2004). tau.ac.il/~elicomay/ LastWord.pdf

further considers the works of Gell-Mann and Ne'eman and of Feynman and Bjorken to be some of the best bodies of work done in decades. Gell-Mann and Ne'eman provided convincing arguments for the existence of quarks as elements accounting for the structure of hadrons, and Feynman's and Bjorken's works established the theoretical basis for the dynamical properties of quarks which have been proven later in experiments.

These theoretical works and the experiments related to them validate quarks' existence. Historically they preceded QCD and are independent of it. Therefore, the assumption that quarks constitute a central element in every theory of the strong interaction seems valid beyond any doubt.

The weak force, which is described in detail in the standard model, will not at all be discussed in this book. We believe that most of the issues regarding the strong interaction can be apprehended without getting into the complexities of the weak force.

The structure of this book

The target readership of this book is mixed. It is targeted to any person who has a good understanding on physics, physics students, and even particle physics experts. The latter may use this book as an easy to read introduction to the corresponding scientific articles.

All the chapters in this book contain new and innovative perspective on issues in particle physics related to the strong interaction, except for the chapter "particle classification," which contains only basic material. Most of the chapters start with some background and continue with deeper discussion.

The next unit of the book, "The shell model of the proton," shows the advantage of explaining strong interaction as a force with similarities to the electromagnetic force.

The third unit, "Additional mass in the proton," discusses phenomena proving that the nucleons must contain massive objects in addition to the three valence quarks and the associated quark-antiquark pairs, and provides some calculations of mass and radius properties of baryons and mesons. The calculations may be a bit difficult for some readers. Readers may skip the sections discussing these calculations.

The fourth unit, "Magnetic monopoles," discusses the theory that Comay developed in the 1980s regarding the physical properties of quarks. This theory constitutes the foundation of the entire model.

The fifth unit, "Summary," starts with an important chapter called "And Yet, Why Do Scientists Believe in QCD?" In this chapter we present the arguments that were provided by particle physics experts in order to convince us that QCD must be correct. These arguments were collected from meetings we had with leading scientists in the field and correspondence with other experts. The last chapters in this part summarize the current situation.

The "Terminology" appendix contains definitions of basic terms for the benefit of people who are not particle physicists. The "Selected Articles of Eliyahu Comay" appendix contains abstracts of some of the relevant articles published during the last 30 years.

UNIT 2: THE SHELL MODEL OF THE PROTON

Chapter 5: Particle Classification

Here we will describe briefly the known particles and classify them according to their properties. Readers who are not familiar with the terminology may refer to the "Terminology" appendix which contains the terms discussed in this book in alphabetical order.

Physicists may skip directly to the tables at the end of this chapter.

Massive and massless particles

According to Special Relativity, a particle that can be at rest cannot move at the speed of light or beyond. Such a particle is at times called "massive". Protons, neutrons, electrons, etc. all belong to this particle category. The photon, on the other hand, cannot exist at rest, and not only that, it has to move at a constant velocity, the speed of light.

This means that particles can be sorted into two categories: those always moving at the speed of light, and those always moving at a speed lower than the speed of light.

In the late 1930s, Eugene Wigner published a fascinating paper known today as "Wigner's Analysis of the Poincaré Group"[88]. He used group theory to reach a similar conclusion. This paper, which

---

[88] E. P. Wigner, *On Unitary Representations of the Inhomogeneous, Lorentz Group*, Annals of Math., **40**, 149 (1939).

was denied publication in a physics journal and was finally published in a mathematical journal, is now considered one of theoretical physics' deepest and most impressive 20th century papers.

Massless particles

In addition to the photon, there are several hypothetical massless particles, such as the "gluons", that many physicists believe carry the strong interaction and the "gravitons" which are believed to be the carrier of the gravitational force.

Massive elementary particles

An elementary particle is a particle that has no internal structure, and a composite particle is a particle composed of several elementary particles. Elementary particles are point-like.

Proving experimentally that a particle is elementary and point-like is not possible. Experiments show that the volume of some particles is smaller than the experimental detection threshold, and they are assumed to be point-like. Several other particles are considered as elementary in spite of the fact that their volume cannot be measured because they decay too quickly.

There are two categories of elementary particles: the "Leptons" and the "Quarks". There is another family of massive particles, consisting of particles which carry forces, that contains the Higgs, W and Z bosons. These particles are not relevant to the main purpose of this book – describing an alternative model to the strong interaction.

The leptons and the quarks obey the Dirac equation, formulated by Dirac in 1928, and they are sometimes called Dirac particles. For every Dirac particle there is an anti-particle that has the same properties but an opposite charge.

Another property which is common for all Dirac particles is their Spin – they all have spin-1/2 which means that their spin is half ℏ. ℏ is the standard unit used for describing the angular momentum (spin) of a particle.

Quarks

There are six types of quarks: Up, Down, Strange, Charm, Bottom and Top. These are denoted with the letter u,d,s,c,b,t respectively. As previously mentioned, for each particle there is an anti-particle, so there are 6 anti-quarks. They are written with a bar (i.e.: $\bar{u}$ is the anti-particle of u).

The quarks have electric charges – they can be +2/3 (u,c,t) or –1/3 (d,s,b). Their anti-particles carry opposite charges: –2/3 and +1/3 respectively.

The quarks participate in Strong Interactions, and they are found only in composite particles, called Hadrons. We will talk about Hadrons later in this chapter.

Leptons

There are two types of Leptons – the Electrons and the Neutrinos. There are 3 types of electrons – electron, muon and tau (or tauon) and for each type of electron there is a corresponding neutrino – electron neutrino, muon neutrino and tau neutrino. Electrons have a charge of -1 and they participate in electromagnetic interactions. Neutrinos have no charge and they do not participate in electromagnetic interactions. All leptons have a corresponding anti-particle.

Quarks, electrons and neutrinos participate in weak interactions and obviously in gravitational interactions. These interactions will not be covered in this book.

61

Elementary particles in one table

Table 2. Elementary particles

Massless particles					
Photon		Gluon*		Graviton**	
Massive Particles					
Leptons		Quarks	Anti-Leptons		Antiquarks
Electron	Electron-neutrino	u,d	Positron	Anti Electron-neutrino	$\bar{u},\bar{d}$
Muon	Muon-neutrino	s,c	Anti-Muon	Anti Muon-neutrino	$\bar{s},\bar{c}$
Tau	Tau-neutrino	b,t	Anti-Tau	Anti Tau-neutrino	$\bar{b},\bar{t}$
Massive Bosons					
W, Z and Higgs bosons***					

* Comay's model does not contain gluons

** Hypothetical particle

*** See remarks in the chapter "The Yukawa and the Higgs Particles"

Composite particles

Atoms and molecules are particles composed of electrons and positively charges atomic nuclei. The force binding electrons to the nucleus is the electromagnetic force. Another composite particle is the positronium, which is composed of an electron and a positron.

The atomic nucleus is also a composite particle composed of protons and neutrons. The force that binds protons and neutrons in the nucleus is called the strong nuclear force. The stable composite of one proton and one neutron is called deuteron.

However in this book we will mainly focus on composite particles which consist of quarks and are bound by the strong interaction. These particles are called Hadrons.

Experiments conducted to date show that there are only two types of hadrons: **baryons** and **mesons**. Other types of composite particles, such as pentaquarks, tetraquarks and dibaryons, were predicted by some physicists but were never found experimentally.

Experiments show that baryons contain 3 valence quarks in their external shell. The proton and the neutron which compose the atomic nucleus are called nucleons, and they are the best known baryons.

Many other baryons exist, but they are not stable and only exist for an extremely short period of time before they transfer to another baryon plus other particles.

There are many quark combinations forming baryons, therefore the baryon family contains many kinds of baryons which have been found in experiments.

The proton is the only baryon which is stable as a free particle. The half-life period of the neutron is approximately 15 minutes, and a free neutron decays into a proton, an electron and an anti-neutrino. Neutrons, however, do not disintegrate when inside a stable atomic nucleus.

Mesons are a composite of a quark and an antiquark. There are many types of mesons because in principle all quarks can form a pair with any antiquark and each pair can make several different mesons.

All mesons are unstable. The most stable mesons are the π^+ and π^- which belong to the group of three pi mesons, or pions. (The superscripts + and − denote the sign of the electric charge of the particles.) The half life period of the π^+ and π^- is roughly 10^{-8} seconds. π^+ is composed of u quark and $\bar{d}$ antiquark, and its antiparticle π^- is composed of $\bar{u}$ and d. The third pion, the π^0 is chargeless and its half life period is approximately 10^{-16} seconds.

Table 3. Hadrons.

Hadrons		
Baryons		**Mesons**
Nucleons	**Unstable baryons**	
Proton, neutron	Numerous types of unstable baryons	Numerous types of mesons

Chapter 6: What Theories Say

The QCD theory is the part of the standard model explaining the strong interaction. The birth of the QCD theory in the late sixties and early seventies of the last century and its establishment as an unshakeable theory is the outcome of a fascinating historical evolution.

In the 1960s and 1970s, scientists had already discovered that hadrons were composed of quarks. When physicists tried to understand the nature of the force holding the quarks together inside hadrons, they had to face quite a few issues, including the following fundamental questions:

- Why do nucleons contain precisely 3 quarks?

- Why are there no particles with 2 or 4 quarks?

- How come there are mesons composed of a quark and an antiquark?

- What are the characteristics of the force holding quarks together?

- What is the explanation for the baryon conservation law which states that the number of baryons in physical processes always remains constant?

- How can one account for the existence of the Ω^- and Δ^{++} baryons? (This topic will be clarified later.)

- How can one account for the similarity between the strong nuclear force and the van der Waals force?

At that time, it was not yet known that nearly half of the linear momentum of a nucleon moving at a very high speed is carried by quarks. This means that half of the nucleons' mass is actually not in these quarks. In fact, not even one theory assumed the existence of massive objects in addition to the quarks.[89]

If such information had been available to scientists at that time, they may have had less difficulty adopting a theory assuming the existence of massive objects inside the nucleons.

What Comay's theory says

Comay's model assumes that the strong interaction is essentially similar to the electric force, which means that particles with an opposite "strong charge" attract each other, and particles with a similar strong charge repel each other.

According to Comay's model, nucleons have a positively strong-charged core, and the quarks' strong charge is negative. Thus the core attracts the quarks, and the quarks repel each other, in an analogy to the action of the electric force in the atom, where a positively charged nucleus attracts electrons. The nucleon core has 3 strong charge units, and each quark's strong charge is one (negative) unit. This core has zero electric charge.

---

[89] Henry W. Kendall: *Deep Inelastic Scattering*: *Experiments on the proton and the observation of scaling.* Nobel Lecture, December 8, 1990. Reviews of Modern Physics, Vol. **63**, No. 3, (1991).

There is a subtlety to the theory: this core contains additional closed inner quark shells, in analogy to an atom filled by electron shells with three electrons in the external shell. These closed inner quark shells contain equal amount of u and d quarks, and probably no other kinds of quarks.

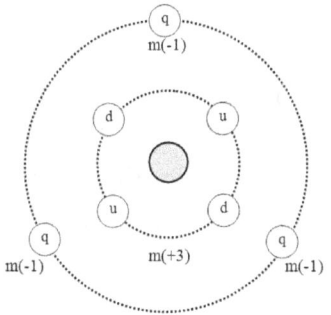

Figure 17. Baryons according to Comay's model. Quarks are attracted to the proton's core because they have a strong charge of opposite sign (like non-ionize atom)

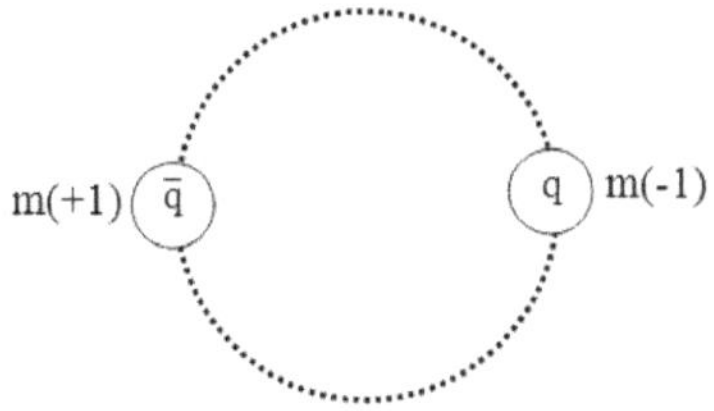

Figure 18. Mesons according to Comay's model. The quark is attracted to antiquark because they have strong charge of opposite sign (like positronium)

The existence of internal quark shells in baryons is an important assumption because it offers an explanation to quite a few phenomena, some of which QCD does not account for. These shells are key to explaining the increase in the elastic (as well as total) cross section curve in higher energy proton-proton collision, the measured mass of the valence quarks of the proton, the radius of the proton and why u,d quarks are heavier in baryons rather than in mesons and other quarks are heavier in mesons.

These issues will be discussed later in this book.

Therefore, without having to generate any major hypothesis of a new kind, the questions physicists are confronted with are readily answered:

67

- The strong charge of the nucleon's core has the same magnitude as the overall charge of the three valence quarks but with an opposite sign. Therefore, the total strong charge of baryons is zero.

- Two quarks repel each other. This is why there are no particles containing only 2 quarks.

- Quarks and antiquarks attract each other and this is the reason for the existence of mesons. The idea that an anti-particle carries a charge opposite in sign to the particle has been an inalienable asset to physics for over a half of a century. Here Comay assumes that the same physical law applies for the strong charge as well.

- The baryon conservation law results from the fact that every baryon has a core, which means that there are as many baryons as the number of cores.

- This model serves as a baseline for understanding the similarity between the van der Waals forces and the strong nuclear force, since both are residual forces of a fundamental electromagnetic-like interaction characterized by the attraction between opposite charges and the repulsion between identical charges.

Comay also developed the quarks' quantum equations (and formulated their Lagrangian density and other properties) which are discussed later in this book.

For now, we will assume that the strong interaction behaves very similarly to the electromagnetic force, although its elementary unit of charge is much stronger.

What QCD says

As opposed to the assumption of the existence of the core inside the nucleons, building the QCD theory required the invention of a long series of new kinds of assumptions, some of which look quite fantastic. In order to account for the key issues around the strong interaction, QCD states that:

- The strong force is associated with a triple charge, called colors (a new idea first presented when QCD was conceived)

- Every quark has a different color

- Only particles containing equal amounts of the 3 colors (of total color "white") can be physically measured. The other combinations are "forbidden" and cannot exist as separate entities.

- The force is carried between the quarks by means of gluons, which are massless particles.

- Every gluon has a non-white combination of color and anti-color (therefore, no individual gluon can be found in a physical measurement).

- Gluons exchange their colors in gluon-quark interactions.

- One of QCD's results is the existence of an attractive force between the quarks, which increases as quarks move away from each other (unlike any other known elementary force in nature).

- At a certain distance, a "cutoff" effect stops the rise of QCD's interaction strength.

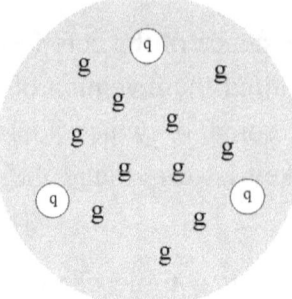 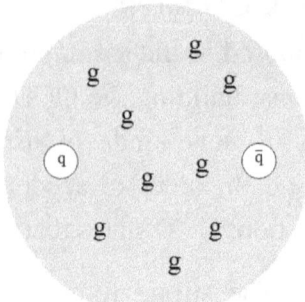

Figure 19. Baryons according to QCD. Quarks are attracted to each other with gluons

Figure 20. Mesons according to QCD. Quark and antiquark are attracted to each other with gluons

After QCD was conceived, it turned out that nearly half of the nucleon mass is not carried by the valence quarks. Therefore, scientists decided that the gluons are the ones carrying this missing mass.

The unexplained strong nuclear force

According to QCD equations, the intensity of the force between 2 quarks grows as they move away from each other. It is therefore unclear why a quark on the moon is not attracted to a quark on earth with a tremendous force. Therefore, QCD assumes that the force stops at a certain distance – this assumption is sometimes called "cutoff" – but for QCD, the explanation for why cutoff occurs is still open.

The question of the strong nuclear force touches both the fields of particle physics and nuclear physics, and there is no single coherent theory admitted by the entire scientific community.

A widespread and justified view qualifies the strong nuclear force as a "residual force", which means that it is the consequence of the action of the strong interaction on the quarks, and this emphasizes its resemblance to the van der Waals force. However, the van der Waals force exists only because the electric force has a simple structure of attraction between opposite charges and repulsion between identical charges. On the other hand, the structure of the force described by QCD is highly complex and does not explain the "residual" nature of the nuclear force in the context of nucleons in the nuclei[90].

Some physicists still hold the explanation provided by Yukawa in 1935 with regard to the strong nuclear force, even though it is not related to any residual force.

In this situation, QCD does not explain why the quarks within one nucleon inside an atomic nucleus are not powerfully attracted to the quarks inside the neighboring nucleon. It is also not clear how the potential curve in Figure 8 is formed. QCD offers no explanation to the fact that the nucleons' density inside heavy atoms is constant[91]. And as we have already mentioned, there is no satisfactory explanation to the first EMC effect.

Comay's model, on the other hand, provides a natural explanation to the phenomena related to the strong nuclear force:

- The strong nuclear force stops acting at a certain distance due to an electromagnetic-like screening effect of the strong charges.

---

[90] S. S. M. Wong, *Introductory Nuclear Physics* (Wiley, New York, 1998). p.102. "Although the idea of a color van der Waals force is pleasing, the actual form it produces has a range much longer than what is observed. Currently, the color van der Waals force does not seem to be a correct model for nuclear interaction without modifications."

[91] Frank Wilczek, *Hard-core revelations*, Nature, Vol **445**, 2007. See discussion later in this book in the chapter "And Yet, Why Do Scientists Believe in QCD?"

- The strong nuclear force is generated in the same way that the van der Waals force is obtained since they both derive from "atom-like" structures of positive charge at the center that attracts negative charges.

- In an analogy to the fact that the volume of atomic and molecular electrons is larger inside liquid, the first EMC effect is explained similarly.

- This explains the similarity of the van der Waals potential graph and the strong nuclear force potential graph.

- The constant density of nucleons in nuclei is explained in the same way as the constant density of molecules in liquid. They both reach an equilibrium state between attraction forces (van der Waals force for atoms or molecules and strong nuclear force for nucleons) and repulsive forces (that stem from the Pauli exclusion principle).

A note of "mathematical" nature regarding the van der Waals force and the strong nuclear force: the most prominent feature of the van der Waals force and the strong nuclear force is the fact that when the particles move away from each other the force rapidly decreases. In the case of the van der Waals force, the phenomena is explained by the "Screening Effect", i.e., when looking at the atom from a distance, electrons and protons carrying opposite charges cancel each other when the electron distribution is spherical, and nearly cancel each other when the electron distribution is not spherical. This effect has a profound mathematical reason: the intensity of the original force decreases according to the famous inverse square law.

Comay's model indeed describes two forces: repulsion between quarks and attraction between each quark and the nucleon's core.

72

These forces are equivalent to electric forces, and also decrease according to the inverse square law. On the other hand, in the case of QCD, which is built of different equations, there is no analogy to the van der Waals force.

The A1 fiasco

Comay completed his M.Sc. degree at the Hebrew University of Jerusalem in the 1960s. During his M.Sc. study he acquired expertise in calculations using the Wigner and Racah theory. He started his Ph.D. work in particle physics at Tel Aviv University, in 1970.

At that time, Comay had already realized that the electromagnetic force and the strong force are very similar.[92] Comay was also aware of the close resemblance between the potential graphs of the van-der-Waals force and the strong nuclear force. On this basis he was convinced that baryonic structure must be similar to the atomic structure. This theme plays an important role in the present book.

Comay suggested to his Ph.D. supervisor that he would like to investigate meson masses for his dissertation topic. Comay thought that his idea was very promising: it took the simplest set of particles and tried to solve the first approximation problem of the state, namely the two-particle Dirac equation. The free parameters were

---

[92] Comay's arguments in 1970 were the following: both forces conserve parity and flavor. Thus, processes determined by these forces are characterized by transitions between energy states of the same constituents (pair production is allowed). Moreover, the elementary constituents of the correspondent systems are spin 1/2 particles, electrons and quarks, respectively. The simplest set of strongly interacting particles is mesons having isospin=1, because they consist only of the two very similar u,d quark-antiquark pair. Hence, these mesons are analogous to quantum mechanical states of the positronium.

quark self-mass and spatially dependent phenomenological interaction. The latter would hopefully take a simple form.

This proposal was presented at the time when the dynamical properties of strong interactions were considered unclear and before the final formulation of QCD. This idea had the potential to yield two kinds of useful information: the phenomenological form of the interaction might provide a clue to the actual form of strong interactions, and the output of the program will provide predictions for the mass of yet unknown isospin=1 mesons and thereby would help future data analysis.

Members of the particle physics group who examined this Ph.D. proposal were not very excited about it. In particular, they doubted the usefulness of the first approximation approach and the fact that field theory effects were ignored. Another objection was based on the unclear statistical properties of quarks. Comay's supervisor asked how he could explain the low mass of Ω^- and Δ^- baryons.

Comay tried to explain how the Wigner and Racah theory could be applied to quarks and account for these particles, but a serious discussion of this issue did not follow.[93]

His supervisor finally allowed Comay to develop his idea – on his own, adding that if Comay managed to publish even one article on the subject, then he would consider joining Comay's efforts. Comay started to build the mathematical model that would be able to confirm the masses of several known mesons. He considered a group

---

[93] Years later, the creation of QCD and the long duration of the proton spin "crisis" convinced Comay that members of the particle physics community simply do not understand the configuration structure of quantum mechanical bound states of several particles. This piece of knowledge is easily obtained from the Wigner-Racah mathematical tools.

of seven known mesons, all of which are made of *u,d* quarks only. Since he was also a good software developer, he received preliminary results quite shortly. In his computerized model he could define parameters for the *u,d* masses and for the potential formula.

He continued by checking whether an electromagnetic-like interaction formula fit the experimentally measured meson masses. Later he tried other kinds of interaction formulas. After a while he made two principal discoveries. Minor modifications of the interaction formula had little effect on the quality of the fit of the calculated values to the experimental data. A second point was that the masses of all mesons could be confirmed by his new model with a very satisfactory error value.

Except for one meson, called A1.

The measured A1 mass was 1070 MeV, but according to Comay's calculations the A1 mass had to be much larger than that. The deviation was rather large. He tried to build alternative models (with different kinds of potential formulas) during the next two years, but without success. Finally, in 1972, he abandoned this research and completed his Ph.D. in another area of nuclear physics.

In 1980, scientists were able to measure the A1 mass more precisely, and found that it was 1260, and not 1070, as believed in 1970 (see Figure 21).

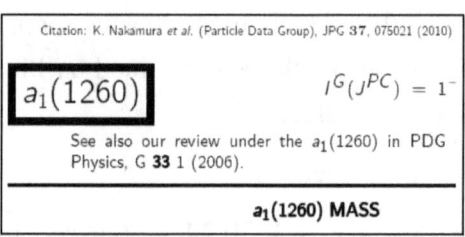

Figure 21. The measured mass of the A1 meson in 1970 (top) and its current value (bottom)

It can be concluded that the idea was abandoned for two direct reasons. One was the incorrect experimental mass value that had acquired official recognition. The second reason was Comay's lack of experience. Somebody else might drop the troublesome A1 meson from the data set and obtain a good fit to the other mesons. On the basis of this fit it is possible to derive a *correct prediction* for the A1 mass.

Nucleon calculation is a further development along this course. Here the similarity between the van-der-Waals force and the strong nuclear force indicates that nucleons have a core. Thus, according to our knowledge, in the early 1970s this was the only theoretical idea requiring that beside quarks, baryons contain another kind of massive elements. This type of proton structure can be used for making a *prediction* stating that quarks do not carry all the proton mass, as it was found several years later.

Comay continued to wonder about the nature of strong interactions. Requiring simplicity, he thought that strong interactions should take an electromagnetic-like form. However, if this is the rule of the game then where are the photons of this kind of interaction? The solution to this dilemma will be found in a quite unexpected place.

Chapter 7: Wigner, Racah and the Three Quarks

During the 1940s, Wigner and Racah separately developed a mathematical theory which today is called "Angular Momentum Algebra" or the "Wigner-Racah Calculus". The theory greatly simplifies the calculation of the electron structure in atoms, but its profound meaning is a potentially major contribution to the understanding of the structure of nucleons.

Based on historical evidence, it would be plausible to assume that if elementary particle scientists were familiar with this theory and had reached a real understanding of its consequences, QCD would have never been conceived. Racah died in an accident in 1965 during a family visit in Italy, and Wigner never really dived deep down into particle theory during the late 1960s, and stopped publishing in this field. The sad truth is that there are perhaps no elementary particle scientists who really understand the theory's profound meaning. Comay recalls meeting a colleague a few years back who told him how, when working on a paper describing Racah's work, he sent it to a major journal, and the editor sent it back saying that there was no scientist capable of reviewing the article, and that none of the scientists associated with the journal had any idea what it was about... The paper ended up being published in a different journal.

Before going deeper into the details in order to understand the theory and its consequences on the understanding of matter, let us look at a few experimental results which are easily accounted in the light of Wigner-Racah's theory:

- The relatively low energy of the particles Ω^- and Δ^{++}, which was one of the major incentives for the invention of QCD.

- The phenomenon called "the second EMC effect" revealing that the total spin of the proton quarks is close to zero. This effect is also called the "proton spin crisis" and was subject to hundreds of articles.

God plays with many dice

One of the most astonishing aspects of quantum mechanics is the use of statistics to describe a particle's state. From the beginning, quantum theory stated that a particle's location cannot be well determined.

Experiments support the assumption that an elementary quantum particle is a point-like particle. However, the theory's founders assumed that a particle is present at different probabilities in many different places at the same time. This concept was so bewildering, that Einstein complained that "God does not play dice". Einstein and his colleagues proposed an experiment, today called the Einstein-Podolsky-Rosen-Bohm experiment, the outcome of which was expected to invalidate quantum theory. When the experiment was finally carried out decades later in 1982 in Orsay, France, the results confirmed quantum mechanics' prediction rather than Einstein's intuitive reasoning[94]... It turns out that the laws of nature do indeed like playing with dice much more than we can grasp.

Additional pairs of particle-antiparticle

In the late 1930s and 1940s scientists took the concept of uncertainty one step further, measuring a free electron, i.e., an electron not located in an atomic shell. It turns out that Dirac particles can be in several states at the same time, (i.e., that the particle's wave function is a superposition of different simultaneous states each characterized by a different probability). Some of these states are described as a superposition of a particle and an additional particle-antiparticle pair. The electron, for example, can be described both as a single electron, and as a superposition of 2 electrons and one positron, etc.

Measurements of the hydrogen atom's energy levels conducted in 1947 by Willis Lamb and Robert Retherford showed that those energy levels were slightly shifted in comparison to the predictions of Dirac's equation for a single electron. This result refuted the single particle approach and supported the theory called quantum electrodynamics. The measured shift is called "Lamb Shift", and its relative magnitude is very small: about 0.000001. This is because when describing the particle as a superposition of many different states, every state has a certain probability to exist, and for the electrons, as Lamb and Retherford discovered, the probability of states with additional electron-positron is very small. A precise

---

[94] A. Aspect, P. Grangier and G. Roger, *Experimental Realization of Einstein-Podolsky-Rosen-Bohm Gedanken experiment: A New Violation of Bell's Inequalities,* Phys. Rev. Lett., **49**, 91 (1982).

measurement of the electron's magnetic moment leads to the same conclusions[95].

In hadrons the strong interaction is much stronger than its electric counterpart, and the existence of additional quark-antiquark pairs, as predicted by quantum field theory, is easier to notice. And indeed, experiments successfully showed that there are antiquarks inside the proton.

Theoretically, the existence of these antiquarks derives from the addition of one or several pairs of quark-antiquark to the proton's wave function. Measurements show that there is in average 0.5 quark-antiquark pairs in the proton[96].

The theory of Wigner and Racah

Calculations based on Wigner and Racah's algebra brought out another aspect of the uncertainty of the electron's state. The electrons in the atomic shells have different simultaneous mathematical forms called configurations, meaning that they occupy several shells simultaneously. All the combinations permitted by the basic laws of physics exist, and each with its own probability![97]

Wigner and Racah developed mathematical tools to calculate the probability of each configuration. The theory they developed for calculating these probabilities is called "Angular Momentum

---

[95] K. Nakamura *et al.* (Particle Data Group), J. Phys. G **37**, 075021 (2010). pdg.lbl.gov/2010/listings/rpp2010-list-electron.pdf (p. 2.)

[96] D. H. Perkins, *Introduction to High Energy Physics*, (Addison-Wesley, Menlo Park, CA, 1987). p. 281

[97] A. W. Weiss, *Configuration Interaction in Simple Atomic Systems*, Phys. Rev. **122**, 1826 (1961).

81

Algebra". Computer calculations carried out in the late 1950s and early 1960s showed that for quite a good number of configurations, probability is not as small as in the case of the single electron in the hydrogen atom, where each state has a unique configuration. For example, if we take the ground state of the Helium atom, naively considered as simple since it only has one electron shell with two electrons in it, 35 (!) distinct configurations exist simultaneously, some of which having a two-digit probability. Incidentally, the configuration number can be brought down to ten[98], but even ten is a large and surprising number for the ground state of a "simple" atom containing only two electrons.

When high speed computers were later built, Wigner-Racah techniques were used to calculate states of multi-electron atoms.

Although this is not one of the hot topics at the front of science, everything that we discussed in this chapter to this point is documented.

However the impact of these physical properties on the understanding of the behavior of quarks in the nucleons is even more farfetched.

Wigner, Racah and the Quarks

As previously mentioned, Comay's model assumes that baryons have an inner core around which the quarks are arranged in shells filling up energy levels inside the nucleons just as electrons do in atoms. If this model is correct, then the same rules that govern electrons in atoms should apply to quarks in nucleons. In fact, the configuration approach holds such a fundamental place in physics

---

[98] It is possible to group several configurations used in Weiss article into one and reduce the number of significant configurations to 10

that they probably exist even without Comay's model. The number of valence quarks in baryons is 3, whereas the number of electrons in the Helium atom is 2. The quarks are subject to forces a hundred times stronger than the electric forces, and just like atomic electrons, quark states are also described as a superposition of many possible simultaneous configurations. In the case of quarks, it would actually be impossible to define one dominant configuration with a significantly higher probability.

The situation is actually even more complicated than that: as opposed to electrons in the atom, where the probability for the existence of additional electron-positron pairs is very small, in nucleons, the probability of finding additional pairs of quarks and antiquarks occupying energy levels is significant: nearly 0.5 pairs of quark-antiquark within the nucleon.

I suppose that most of the readers, who tried to follow up to here, went out for a cigarette break, even if they are non smokers... but let's continue anyway.

The proton spin crisis

If we sum up the spins of all these different quark configurations inside the proton, the total will be close to zero. Why? Because adding up spins is a directional addition (vectorial addition, for those who are familiar with this term), which means that spins in opposite directions cancel each other.

Since the spins of the quarks in the different configurations sometimes cancel each other, the total nucleon spin will be close to zero, and this is, as we mentioned earlier, the problem called the "proton spin crisis".

If elementary particle scientists were familiar with this demonstrated physical knowledge, the proton spin crisis would not have turned into a crisis at all. Unfortunately, the proton spin crisis is still considered an unresolved problem[99].

The Ω^- and the Δ^{++} baryons

The unexpected properties of the Ω^- and Δ^{++} baryons, discovered experimentally, led the QCD pioneers to look for an out of the box solution and resulted in further developments of the QCD theory. Δ^{++} and Ω^- are baryons, specifically, particles from the nucleon family, which have three u and three s quarks respectively and spin-parity $3/2^+$.

The energy of these baryons is relatively low. Because of the Pauli exclusion principle, it is impossible to come up with a total spin-parity of $3/2^+$ and isospin $3/2$ from a three quarks combination, based on a single configuration in which all quarks are in the spatial state of minimal kinetic energy. Scientists were embarrassed by this only because they considered a unique particle configuration, that is, they assumed the particle had only one definite state.

To consider the particle as a superposition of several simultaneous configurations is a significant step toward resolving the enigma.

The following part of the explanation about the Δ^{++}, published a few years ago[100,101], is intended for physicists.

---

[99] Wikipedia list of unsolved problems in physics

[100] E. Comay, in "Has the Last Word Been Said on Classical Electrodynamics?" ed. A. Chubykalo, V. Onoochin, A. Espinoza and R. Smirnov-Rueda (Rinton Press, Paramus, NJ, 2004).

84

An important notion for the understanding of the structure of Δ^{++} is what is called in atomic spectroscopy "exchange integrals" (or "exchange term"[102]). Exchange integrals appear explicitly in a configuration of electrons (or quarks here) located in different shells. It turns out that the exchange integral is very fond of states in which spins are symmetric (i.e. parallel) and the spatial function is anti-symmetric[103]. The spin related Hund's rule for atomic energy levels which says that ground states favor parallel spins is related to this issue.

In general, parallel spins lead to an anti-symmetric spatial part of the total wave function which increases the binding energy, as is formulated in one of Hund's rules for atomic spectroscopy. This property can be illustrated in the following picturesque manner: electrons (or quarks here) in an anti-symmetric spatial wave function component are farther apart from each other. Therefore the positive repulsion energy is smaller and the binding energy is larger. This means that the combination of a symmetric spin and anti-symmetric spatial functions is more frequent in low energy states (such as the baryon Δ^{++} or in atoms obeying Hund's rule).

---

[101] For a complete explanation about Δ^{++}, Ω^- and the proton spin crisis see E. Comay, *On the Quantum Mechanical State of the Δ^{++} baryon*, Prog. In Phys. **1**, 75 (2011). tau.ac.il/~elicomay/dpp_fin.pdf.

[102] A. de-Shalit and I. Talmi, Nuclear Shell Theory (Academic Press, New York, 1963). Chapter 20.

[103] Take for example the pairs of singlet and triplet of states of the He atom. Here in a triplet S=1 state, the spins are parallel and symmetric whereas the spatial components of the wave function are anti-symmetric. In the singlet S=0 states, these relations are reversed. Thus, due to the exchange integral, a triplet S=1 state is bound stronger than the corresponding singlet S=0 state.

The following point emphasizes the importance of the previous argument. Based on the proton's dimension and Heisenberg uncertainty principle, the energy necessary for the spatial excitation of a quark in a proton can be estimated to some few hundreds of MeV. On the other hand, excited baryonic states show that the binding energy is much larger than 1000MeV. For comparison purposes, in the hydrogen atom, it is well known that the ratio between the kinetic and potential energy is one half.

Therefore, quarks in the proton would find it "really cheap" to "pay" with kinetic energy and save the repulsion energy between the quarks. These considerations provide a qualitative explanation to the relatively low energy level of the Δ^{++} baryon.

One more comment

The impact of Wigner and Racah's arguments goes even further. Regarding the significant probability for the existence of additional quark-antiquark pairs in the nucleon, any calculation of quark energy states in nucleons should take these pairs and the multiple configurations into account, because, unlike the electron in the atom – the multiplicity of additional quark configurations does have a more significant[104] impact on the calculations of physical properties such as mass, magnetic moment etc.

Incidentally, it seems like most (if not all) particle physicists are unaware, even today, of the necessity of using multiple configurations for describing quark states. Indeed, the "proton spin

---

[104] Indeed, as is well known from the Hamiltonian's hermiticity, the higher multiplicity of states increases the probability of finding the lowest state at a lower energy.

crisis" has been going on for decades and is not explained even today.

Chapter 8: Repulsive Forces inside the Proton

The significant existence of quark-antiquark pairs inside the nucleons[105] is the source of several interesting phenomena.

The antiquark is located in the proton's periphery

Measurements show that antiquarks tend to be located in the proton's external regions whereas quarks settle in a smaller volume inside the proton[106].

Why is that?

According to Comay's model, there is an inner core inside each nucleon characterized by a positive strong charge. The nucleon's valence quarks carry a negative strong charge and are therefore attracted to the core. The antiquarks, on the other hand, which carry a positive strong charge, are pushed away from the core. This effect explains the antiquark presence in the external regions of the nucleon.

---

[105] D. H. Perkins, *Introduction to High Energy Physics*, (Addison-Wesley, Menlo Park, CA, 1987). p.281

[106] Refer to the graph in the previous reference of Perkins. The x-width of antiquarks is smaller, and therefore has smaller Fermi motion and larger 3D volume.

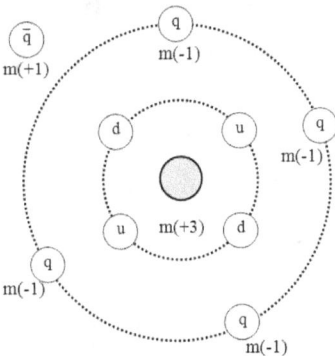

Figure 22. A simplified illustration showing why antiquarks tend to be located in the nucleon's periphery (see text).

QCD provides no explanation for this phenomenon. In fact, this phenomenon contradicts QCD.

The pi meson (called pion) is composed of a bound pair of u,d quark-antiquark. These quarks are of the same kind as those found inside the proton. The pion radius is somewhat smaller than that of the proton. This means that the attractive force of a single quark can confine an antiquark in a sphere which is smaller than that of the proton.

On the other hand, considering the proton component where an antiquark exists, we have four quarks (the three valence quarks and the antiquark's companion). It is unclear why four quarks are not able to confine the antiquark at least within the proton's space in which they reside, and allow the antiquark to occupy a larger volume. If QCD were correct then the antiquark's volume would be much smaller than the measured quantity.

This QCD problem becomes even more obvious, because here the Pauli Exclusion Principle does not impose any spatial constraint on a single antiquark.

89

Electric charge distribution within the neutron

Another interesting phenomenon is the propensity of negative electric charges inside the neutron to be located in more peripheral areas than the positive charges[107]. The neutron contains a u-quark with a charge of 2/3, and two d-quarks, each with a charge of -1/3. As in the proton, there are other quark-antiquark pairs inside the neutron[108].

Experiments show that neutrons contain more $u\bar{u}$ -quark-antiquark pairs than $d\bar{d}$ -quark-antiquark pairs[109]. This is explained by the Pauli exclusion principle, stating that identical quarks cannot occupy the same quantum state. Since there are already two d quarks as opposed to a single u quark in the neutron, the addition of d quarks is less likely. The situation in the proton is the exact opposite: there are more d-quark-antiquark pairs, because the proton contains 2 u and a single d.

[107] K. Nakamura *et al.* (Particle Data Group), J. Phys. G **37**, 075021 (2010). pdg.lbl.gov/2010/listings/rpp2010-list-n.pdf p. 4.

[108] Derives from isospin symmetry

[109] M. Alberg, *Parton distributions in hadrons,* Prog. Part. Nucl. Phys., **61**, 140 (2008).

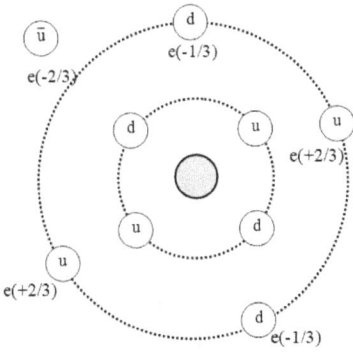

Figure 23. A simplified illustration showing why a negative charge tends to be located in the neutron's periphery. The $\bar{u}$ has the same sign of "strong charge" as the core and is repelled to the neutron's periphery.

As we mentioned above, the antiquarks within the nucleon are pushed toward more peripheral areas. Thus $\bar{u}$ is located toward the external zone. Since the $\bar{u}$-antiquark has a negative electric charge of -2/3, the negative charge of the neutron with $u\bar{u}$ pair will be located in the neutron external region. And since $d\bar{d}$ has a lower probability to be found in the neutron, and the absolute charge value of $\bar{d}$ is smaller (1/3) than that of $\bar{u}$, it does not cancel out the contribution of the negative charge of $\bar{u}$.

QCD has no explanation for this phenomenon either.

Chapter 9: Confinement and Asymptotic Freedom

Some phenomena are successfully explained by QCD. Let us point out the most famous of them, which are considered proof of the veracity of the QCD Theory.

Confinement and asymptotic freedom

When protons are bombarded with high energy electron beams, two seemingly contradicting phenomena occur. On one hand, the quarks inside the proton seem free, that is, their binding energy is very small compared to the energy of a highly energetic incident electron. One could therefore reasonably assume that if the incident electron's energy is high enough, it could tear the quark off the proton. But experiments show that a single quark cannot be torn off the proton, even with highly energetic beams.

The QCD notions of confinement and asymptotic freedom do indeed explain both phenomena, by suggesting that the attraction forces between the quarks increase as the distance between them increases. This concept is contrary to the behavior of the natural laws governing other forces in physics as we know them, that is, that the intensity of the force decreases with the distance between the objects.

The phenomenon of the quark's relative freedom with regard to an incident particle is equivalent to the atomic "Compton Scattering", in which a high energy photon hits an atomic electron. It is known that a highly energetic photon is capable of tearing an electron off (i.e. ionizing the atom).

What causes asymptotic freedom and why is it impossible to tear the quark off the proton?

Comay's explanation to asymptotic freedom property

In atoms with one shell such as hydrogen or helium, the magnitude of the attraction force applied to electrons decreases while the distance between the electrons and the atomic nucleus increases. This is a well known result of the electromagnetic force equations. In larger atoms, the electrons in the external shell are attracted to the positive nucleus. However, due to Pauli exclusion principle they are also repelled from the electrons in the inner shells. Therefore, external shell electrons are bound better when they are located not too close to the atomic nucleus.

Something similar happens in Comay's proton model. The valence quarks are of the u,d flavor, like the quarks in inner shells, and are therefore repelled from inner proton quark shells due to Pauli exclusion principle.

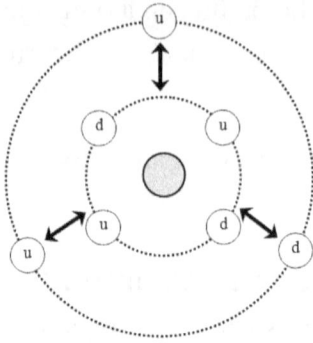

Figure 24. Proton according to Comay's model. The 3 valence quarks are repelled
from the inner quark shells due to Pauli exclusion principle

Incidentally, according to Comay's model this result does not appear
in mesons that have neither a core nor inner shells, or with baryons
with valence quarks which do not have the *u,d* flavor.

In fact, there is strong evidence that asymptotic freedom does not
exist in the general case, as we can see by observing the proton-
proton cross section graph.

The P-P cross section graph vs. asymptotic freedom

The cross section graph is measured in experiments where two
particles collide. This graph provides useful information about the
forces between two interacting particles and about the distance
dependence of these forces.

According to quantum mechanics a particle behaves as a wave
whose properties depend on the particle's velocity. The wavelength
of a particle is shorter when the particle moves faster. Therefore,
when two particles collide, the effective distance between them at
the collision time cannot be much smaller than their wavelength.

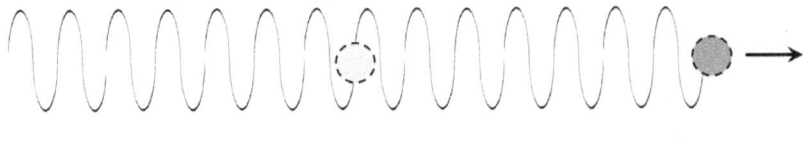

Figure 25. The effective distance between two particles at the time of their collision depends on their wavelengths. Their wavelengths depend on their velocity.

What happens when two particles hit one another? This question is not as trivial as it sounds. When particles do not apply forces on each other, they may go through one another without any interaction. Neutrinos, for example, pass through the entire earth easily. A collision appears when the particles interact and momentum is exchanged.

We can learn about the electromagnetic forces by examining the formula that describes the measured cross section of electron-proton collisions. When an electron hits a proton, their interaction stems only from their electric charges. The cross section curve decreases when the energy of the electron increases. Here the slope of the cross section curve is related to the potential of the electromagnetic field. According to the electromagnetic equations, the potential is proportional to $1/r$ where r denotes the distance from the charged particle to the point where the potential is measured. Calculations show that for high energies, this potential formula yields a cross section graph of the electromagnetic interaction that decreases like $1/p^2$, where p denotes the particle's momentum.[110]

It is convenient to use logarithmic scales in cross section graphs. Thus, the $1/p^2$ relation looks like a straight line.

Now, after having some ideas about cross section curves, let's examine the proton-proton cross section graph.[111]

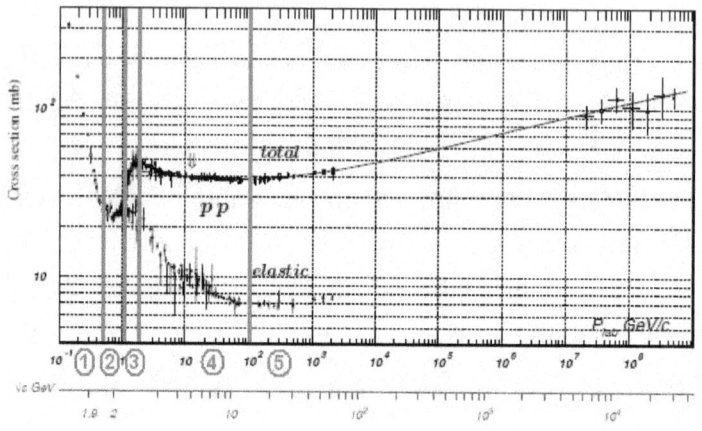

Figure 26. Proton-proton cross section curve, divided to five zones.

Here we add red vertical lines in order to divide the graph into five zones. The zones' order indicates an increase of the collision energy and a decrease of the corresponding proton's wavelength.

In the leftmost zone, zone 1, the proton's energy is rather low and its wavelength is quite long. Here the effective distance between the particles during their collision is so long, that the actual force involved is the electromagnetic force associated with the protons'

[110] D. H. Perkins, *Introduction to High Energy Physics*, (Addison-Wesley, Menlo Park CA, 1987). p. 186-189.

[111] C. Amsler *et al.* (Particle Data Group) Physics Letters B667, 1 (2008) see p.12 in pdg.lbl.gov/2009/reviews/rpp2009-rev-cross-section-plots.pdf

overall electric charge. We can see that in this zone the graph behaves as expected: it decreases in a straight line, proportional to $1/p^2$.

In zone 2, we can see that the graph stops decreasing. In this zone the proton's momentum is larger than in zone 1 and the corresponding wavelength is shorter. In this zone, the effective collision distance is quite short and the protons apply strong nuclear force on one another. This force varies rapidly and the $1/r$ of the potential formula does not hold any more. This is the reason for the rapid change in the slope of the cross section graph.

In zone 3, the graph splits into two parts: the elastic cross section (the lower graph) and the total cross section (the upper graph). We will discuss these two graphs later in the chapter "Something Else inside the Proton." For now, we should know that the two graphs satisfy the rules that we mentioned earlier.

In zone 3, the wavelength is shorter and the protons are so close during their collision that in some cases individual quarks of one proton interact with quarks of the other. This means that the strong force becomes significant. This force is much stronger than the strong nuclear force and it explains the rapid increase of the total cross section graph.

Up to here everything is agreed by all particle physicists.

Zone 4 represents higher energy collisions and shorter proton's wavelength. Here individual quarks of one proton collide with quarks of the other. This part of the graph is relevant to our present discussion.

The behavior of the graph at zone 4 shows beyond any doubt that there is no asymptotic freedom.

How do we expect the graph in zone 4 to behave if asymptotic freedom holds?

If the strong force would behave similarly to the electromagnetic force, meaning that the force increases while the distance decreases, and the increasing potential is proportional to $1/r$, then the graph would drop similarly to its behavior at zone 1. However, if one adopts asymptotic freedom, then the force decreases while the distance decreases. This implies an even more dramatic declining slope of the cross section graph. However, contrary to the asymptotic freedom expectation, the graph's decreasing slope is even less steep than that of zone 1.

The outcome of this discussion is that inside the proton, the force between the quarks of one proton and the quarks of the other increases while the distance decreases, contrary to QCD's asymptotic freedom expectation.[112]

According to Comay's model, the strong force potential behaves like the electromagnetic force. The reason that the graph decreases more slowly than in zone 1 is due to the screening effect of electromagnetic-like interaction. In zone 5, both elastic and total cross sections increase and QCD's asymptotic freedom failure becomes even more evident. This topic is discussed in the chapter "Something Else inside the Proton."

---

[112] The contradiction between QCD and the total cross section graph was mentioned more than a decade ago. A. A. Arkhipov, *On global structure of hadronic total cross sections*, 1999. arxiv.org/PS_cache/ hep-ph/pdf/9911/ 9911533v2.pdf

Another asymptotic freedom contradiction

Physicists use several methods for finding information on the density distribution of components of a composite particle. For example, in the case of the hydrogen atom we have a very successful theory called quantum mechanics. This theory proves that the electron is found in a higher probability near the atomic center, rather than in the periphery. This is explained by the fact that any stable physical system tries to reach its lowest energy state, and the electromagnetic potential decreases while the distance between the electron and the hydrogen nucleus increases. Hence, the density of the negatively charged electron is greater near the atomic center.

In the case of the proton, which is a composite particle as well, it is possible to analyze measurements and build a function called "form factor" that can be used for deriving the proton's quark density. This experimentally based method is independent of any specific theory that aims to describe quark dynamics.

The electric form factor of the proton was measured many years ago. Physicists found that the behavior of proton's charge density that was derived from this function is very similar to that of the hydrogen atom.[113] It can be concluded that the distribution of the charges inside the proton is similar to what we know about the distribution of the electron inside the hydrogen atom.

The same result was achieved when physicists measured the magnetic dipole distribution inside the proton. Since the magnetic dipole stems from the quark charges, then it can be concluded that both charge and magnetic dipole density represent quark density.

---

[113] Using a mathematical terminology, it was found that in both cases the density decreases exponentially with the increase of the distance from the center.

Therefore, the distribution of the quarks in a proton is very similar to the distribution of the electron in the hydrogen atom.[114]

The similarity between the density distribution of the electron in the hydrogen atom and that of the proton's quarks indicates that the decrease of the force derived from the strong interaction is analogous to that of the electromagnetic force, where the potential is proportional to 1/r.

This fundamental property of the proton's form factor is known for several decades and provides another contradiction to QCD's asymptotic freedom because QCD claims that the strong force *increases* with the increase of the inter-quark distance.

Comay's explanation to quark confinement

The explanation suggested by Comay for the confinement phenomenon is based on well-known facts. It has long been established that when an energetic electron beam hits the proton, the outcome is almost exclusively inelastic collisions, that is, interactions in which new particles are created.

It is also well established that when a particle encounters its antiparticle, they both annihilate and release an amount of energy equal to the total of their masses. More precisely, the energy released in particle-antiparticle annihilation is equal to the sum of the particles' masses at the time of annihilation, which can be bigger than their rest mass if they have kinetic energy, or smaller than their rest mass if they are in a bound state and have binding energy.

---

[114] D. H. Perkins, *Introduction to High Energy Physics* (Addison-Wesley, Menlo Park, CA, 1987). p. 194-196

The reverse process occurs similarly. In a collision event, a highly energetic particle can lead to the creation of a particle-antiparticle pair, reducing its energy by the amount of the energy of the newly created system. When an energetic particle hits a proton, mesons are usually produced, that is, a bound quark-antiquark pair.

Pair creation and annihilation is a well known phenomenon in physics. In the case of the electron-positron pair, the binding energy is very low compared to their mass, and the process of pair creation usually yields a free electron and a free positron. For quarks, binding energy is very high and consumes most of its self-mass[115], leading to the production of bound pairs, i.e. mesons.

What goes on inside a proton bombarded by electrons?

When the quark inside the proton gets hit and gains a large amount of energy, the collision usually yields creations of mesons each composed of a quark-antiquark pair, reducing the energy of the target quark. If a single quark could be torn off the proton, its energy would be in the order of thousands of MeV. But 140 MeV would be sufficient to create a meson.

Nature always chooses the least energy path, and therefore, the target quark ends up losing its energy into pair creation, rather than escape from the proton. When a quark-antiquark pair is created, the energetic quark can associate to the antiquark of the pair. Furthermore, we know that the proton contains quark-antiquark pairs, and this means that a quark hit by an electron can annex an antiquark, become a meson and freely exit of the proton.

---

[115] The self-mass of a particle is its mass when it is free and at rest. Even though quarks are always in bound state they have a physical self-mass.

Therefore, the quark does not escape the proton, because it is much easier for a single quark to create a particle pair and join an antiquark in its vicinity within the proton. Together with the antiquark, the combined pair becomes neutral with respect to strong charge and it is practically unaffected by the huge potential barrier created by the core's attractive strong field. Therefore, the new quark-antiquark pair freely escapes the proton.

Comay's explanation of the confinement phenomenon does not require additional assumption (like the "cutoff" of the QCD attractive force) and it flows naturally from his model.

The advantage of Comay's explanation – why mesons can escape nucleons

QCD does not provide a clear explanation why mesons created in energetic collisions can overcome "confinement" and exit the proton. As a matter of fact, this problem is closely related to another one of QCD's failures – its prediction of a baryon-meson strongly bound system, called pentaquark. We will talk about pentaquarks later in this book.

Comay's explanation of the fact that mesons can escape the proton is simple: the quark and antiquark composing the meson carry opposite strong charges which cancel each other out, leaving the meson neutral with regard to the strong interaction. Being neutral, the meson practically does not react to the strong interaction and is free to exit the proton.

One last note regarding this matter: as in the mechanism that creates the van der Waals force, there is a small attractive force between the proton and the meson, but it is residual, meaning that it is much less powerful than the strong interaction. In energetic collisions the newly created meson generally has enough kinetic energy in order to overcome this residual force and exit the proton.

Chapter 10: QCD's Predictions

In the 1960s the understanding was that baryons are characterized by three quarks and mesons by quark and antiquark. A quark combination which does not correspond to the foregoing is considered "exotic". In 2008 PDG[116] published a review of the status of all exotic hadrons[117].

It turns out that QCD sets no absolute constraints prohibiting the existence of exotic hadrons[118]. This explains the abundance of papers based on QCD published in scientific literature, predicting the existence of quite a few exotic hadrons as well as their corresponding nuggets of matter.

Before we get into the details, let us point out a few general statements regarding QCD's framework.

According to QCD, the strong interaction has three kinds of related charges called "colors". A QCD basic structural constraint states that

---

[116] "Particle Data Group", an international collaboration that reviews particle physics and related areas of astrophysics, and compiles/analyzes data on particle properties.

[117] C. Amsler *et al.* (Particle Data Group) *Review of Particle Physics: Pentaquarks*, Physics Letters **B667**, 1 (2008)

[118] en.wikipedia.org/wiki/Exotic_hadron

a free hadron must be "white", meaning it must contain an equal amount of the three colors.

The existence of mesons is explained by the fact that they are composed of a pair of quark and antiquark bound together by an attraction force. In addition, and in order to account for the existence of the proton and the other baryons, QCD suggests that there is also attractive force acting between the quarks.

The abovementioned QCD structural constraint preventing "non-white" hadrons from being experimentally measured limits the possible particle types. But in spite of this rule, QCD still allows the existence of quite a few exotic hadrons as well as various particle nuggets. Some of these particles predicted by QCD are listed below. None of them has ever been found, but in spite of this decades-long failure, the search for them still continues.

Dibaryons

Dibaryons (sometimes called "Hexaquarks") are particles composed of 6 quarks. Their existence was predicted back in 1977 by Robert Jaffe. The existence of dibaryons was considered plausible from the QCD perspective, and some of them are even expected to be stable, especially those composed of the quark combination *uuddss*[119].

According to Comay's model, quarks repel each other, and they are bound to the baryonic core that has strong charge with a magnitude of three units. Therefore, six quarks could only live together within two cores, like in the deuteron. In this case, the force holding these nucleons together is the strong nuclear force. This force is a residual force and is considerably weaker than the strong interactions. Thus, a

---

[119] R.L. Jaffe, *Perhaps a Stable Dihyperon?*. Physical Review Letters **38** 195, (1977).

proton-neutron bound state is the deuteron (heavy hydrogen) and its binding energy is 2.2 MeV, whereas strong interactions are measured by hundreds of MeV. Therefore, a state where two baryons are bound by the strong interaction would be impossible.

Up to now, dibaryons have never been found.

Pentaquarks

Another exotic particle predicted by QCD is the pentaquark. The pentaquark is assumed to be composed of a proton (or a neutron) and a meson bound together by the strong interaction. Their existence was predicted in 1987, and since then hundreds, if not thousands, of theoretical papers have been written about them. Some of these papers even sorted the pentaquarks into several particle categories, and detailed the properties of each of these hypothetical pentaquark particles. In 2003, the LEPS laboratory in Japan announced their discovery, but repeated experiments refuted this finding.

According to Comay, pentaquarks cannot exist.

All hadrons are neutral with regard to their total strong charge. The proton can be compared to a non-ionized atom, and the meson can be compared to a positronium, which is a bound state of an electron and a positron. Therefore, the bond between two hadrons should be comparable to the force between two uncharged particles, as is the case with what is found in the deuteron.

Comay's model predicts therefore that strongly bound pentaquarks will never be found.

2008's PDG[120] report on pentaquarks describes the situation as follows: "... There are two or three recent experiments that find weak evidence for signals near the nominal masses, but there is simply no point in tabulating them in view of the overwhelming evidence that the claimed pentaquarks do not exist..." The report concludes: "The whole story – the discoveries themselves, the tidal wave of papers by theorists and phenomenologists that followed, and the eventual "undiscovery" – is a curious episode in the history of science"[121].

Pentaquarks deserve one more comment. Since QCD claims that all quarks and antiquarks attract each other, physicists concluded that a nucleon and meson composite should be stable. When an energetic particle hits a proton, new mesons (namely, quark-antiquark pairs) are created, and they exit the proton easily. This is in total contradiction to the confinement of the other quarks. Here too, baryon and meson are not bound strongly, as it flows from Comay's model.

The search for pentaquarks is still going on.

Glueballs

Glueballs are particles assumed to be composed singularly of gluons. According to QCD, this quantum state is possible because gluons

---

[120] PDG, the Particle Data Group, is the international organization which is officially authorized to establish the existence of particles and determine their properties.

[121] C.G. Wohl (LBNL), *Pentaquarks*, 2008. pdg.lbl.gov/2009/reviews/ rpp2009-rev-pentaquarks.pdf

contain "color" and operate through the strong interaction. Some combinations of gluons can be "white".

QCD's theoretical calculations predict that glueballs could be formed in the current particle accelerators[122].

According to Comay's Model, there are no gluons and no colors, and there can obviously be no glueballs.

In spite of all the experiments, the existence of glueballs has not been confirmed to date.

Strange Quark Matter

"Strange Quark Matter" (SQM) is the name given to a particular type of hypothetical matter, composed of u, d and s quarks, such as for example the Λ 1116 baryon (which means a combination of 3 uds quarks)[123]. This system is electrically neutral.

The search for nuggets of SQM covered every possible area, including within terrestrial minerals and lunar rocks[124].

According to Comay's model SQM cannot exist. Based on the shell model of the nucleons, binding energy between baryons would be based on a residual force comparable to the strong nuclear force

---

[122] en.wikipedia.org/wiki/Glueball. *"In particle physics, a glueball is a hypothetical composite particle. It solely consists of gluon particles, without valence quarks. Such a state is possible because gluons carry color charge and experience the strong interaction. Glueballs are extremely difficult to identify in particle accelerators, because they mix with ordinary meson states. Theoretical calculations show that glueballs should exist at energy ranges accessible with current collider technology. However, due to the aforementioned difficulty, they have (as of 2010) so far not been observed and identified with certainty."*

[123] E. Witten, *Cosmic separation of phases*, Phys. Rev. **D30**, 272 (1984).

[124] K. Han *et al.*, *Search for Stable Strange Quark Matter in Lunar Soil*, Phys. Rev. Lett. **103**, 092302 (2009).

binding energy. It is known that for a typical nucleus, this energy is about 8 MeV multiplied by the number of nucleons[125].

On the other hand, the energy of the Λ baryon is nearly 177 (1116-939) MeV higher than that of the nucleon. Obviously, the extra 177 MeV energy cannot be stabilized by a force whose strength yields 8 MeV bond. Therefore, since the Λ disintegrates within less than a billionth of a second, SQM is not supposed to exist.

Experimental results are consistent with this conclusion.

Other exotic combinations

Within the QCD framework, many possible combinations of exotic particles are allowed. One of them is called tetraquark, which is supposed to be a stable state of two quarks and two antiquarks.

The existence of none of these hypothetic particles has ever been confirmed. The 2008 PDG review[126] contains a long list of references to scientific literature discussing them.

The efforts to discover this kind of particles continue, including the LHC experiments at CERN.

---

[125] S. S. M. Wong, *Introductory Nuclear Physics* (Wiley, New York, 1998). p.10

[126] C. Amsler *et al.* (Particle Data Group) *Review of Particle Physics: Pentaquarks*. Physics Letters **B667**, 1 (2008)

Chapter 11: The Yukawa and the Higgs Particles

As we mentioned earlier, there is no consistent and agreed-upon theory that explains the strong nuclear force, which holds the nucleons together. It is assumed that this is a residual force of the strong interaction, but the mechanism that creates this residual force is unknown.

Therefore, the old theory formulated during the 1930s by the Japanese scientist Yukawa to explain the strong nuclear force has survived until today.

Yukawa's theory hypothesizes the existence of a massive particle, called the Yukawa particle, carrying the strong nuclear force.

The Higgs Boson is also a massive particle with some similarities to the Yukawa particle. They both belong to a family of particles which, by definition, are massive, elementary and spinless. So far it has been assumed that such particles truly exist, and Yukawa even won the Nobel Prize for allegedly finding a particle from this family.

Based on the knowledge existing today, this kind of particle has clearly not yet been found. This fact has now become common knowledge, as illustrated, for example, by the "Klein–Gordon equation"[127] or the "Higgs Boson"[128] entry in Wikipedia.

Why is the Higgs important?

One of the building blocks underlying the standard model is "the Higgs boson", also called "God's particle" in the popular literature due to its critical role in the standard model.

So far, the Higgs particle has never been observed in experiments in which it was supposed to be discovered. Physicists who stand for the standard model assume that this is due to insufficient experimental energies. To overcome this (and in order to make further discoveries), the most complex scientific device ever built by man – the LHC particle collider at the CERN research laboratories in Switzerland – has recently been completed. This impressive machine is the result of an international endeavor involving over 100 countries.

The non-discovery of the Higgs particle under these ultimate conditions would be a serious blow to the standard model. John Ellis, from CERN, said in 2007: "If you see nothing[129], in some sense then, we theorists have been talking rubbish for the last 35 years[130]." Shlomit Tarem, the head of the Israeli research group in

---

[127] en.wikipedia.org/wiki/Klein–Gordon_equation. *"... no spinless elementary particles have yet been found, although the Higgs boson is theorized to exist as a spin-zero boson, according to the Standard Model."*
[128] en.wikipedia.org/wiki/Higgs_boson. *"At present there are no known elementary scalar bosons (spin-0 particles) in nature, although many composite spin-0 particles are known."*
[129] Not find the Higgs in LHC experiment
[130] The New York Times, 2007. www.nytimes.com/2007/05/15/science /15cern.html

LHC said in 2008: "If we don't find the Higgs then it doesn't exist, and the standard model is not valid."[131]

Are we about to witness the discovery of the first particle belonging to the family of massive spinless elementary particles, as predicted by the standard model?

Can spinless, massive elementary particles exist at all?

Some history

In mid 1920s, Schrödinger tried to find an equation that would describe the electron's quantum behavior. He used the experimentally well-known energy levels of the hydrogen atom to test the validity of his equation. He first tried to use an equation later called "The Klein-Gordon Equation" but realized that it did not yield the desired results. He thus developed the Schrödinger Equation, describing a massive elementary particle.

This equation, published in 1926, was considered a breakthrough, but had a few drawbacks: it did not take the spin into account and was not coherent with the theory of special relativity. In 1927, Pauli published the Pauli Matrices, which extend the scope of the Schrödinger equation and treat the electron as a particle with spin.

A year later Dirac introduced the Dirac Equation, describing an elementary particle which is coherent with special relativity. The Dirac equation gave an excellent description of the electron, which was the only massive elementary particle known at the time. The Dirac equation explains the spin of the electron, as well as the magnetic moment associated to it. Another outcome of this equation is the existence of an anti-particle for each of the particles to which

---

[131] Ynet, 2008. www.ynet.co.il/articles/1,7340,L-3570552,00.html

the equation applies. This result is considered today as an inalienable asset of particle physics.

The Dirac equation minimized Pauli's achievement. It is, therefore, easy to understand why Pauli used to tease Dirac on different occasions. He said about Dirac, who was atheist: "There is no God and Dirac is his prophet"[132], and following Dirac's publication of an article about magnetic monopoles, Pauli named him "Monopoleon"[133].

In 1934, Pauli and Weisskopf revived the Klein-Gordon (KG) equation, describing massive spinless elementary particles. However, spinless elementary particles were not known at that time. When publishing the article, Pauli himself admitted that he didn't find any use for the equation, but that he was happy to have the opportunity to "cast aspersion on my old enemy – the Dirac theory of the spinning electron".[134]

Dirac contested this equation, and published several papers throughout his life arguing and explaining why the KG equation could not be valid. But the scientific community ignored Dirac, whether for relevant reasons, or because they attributed the position he took in his rivalry with Pauli. Since the revival of the KG equation by Pauli and Weisskopf, it has often been used to predict new elementary particles.

---

[132] G. Farmelo, *The Strangest Man* (Basic Books, New York, 2009). p.138.

[133] G. Farmelo, *The Strangest Man* (Basic Books, New York, 2009). p.343.

[134] A. I. Miller *Early Quantum Electrodynamics* (University Press, Cambridge, 1994). p.70.

The Yukawa Particle

In his attempt to explain the strong nuclear force, Yukawa used the idea of a force-carrying particle. According to Yukawa's theory, published in 1935, this particle was elementary, massive and spinless. This kind of particle could be described by the KG equation. Yukawa also provided an estimate of the particle's mass.

We need to remember that in the 1930s physicists believed that protons and neutrons were elementary. They noticed that the force between two nucleons drops down very quickly, but they couldn't jump to the conclusion that such force may be residual force of a stronger force that holds the nucleon's constituents together. The KG equation looked like something that could work, because a solution of the wave equation of a KG particle decreases rapidly and vanishes at a very short distance.

In 1947 a spinless particle was discovered, whose mass corresponded to Yukawa's prediction, and Yukawa won the Nobel Prize in 1949. But after the discovery of the quarks, it turned out that this particle, the π^0 meson, is composed of quark and anti-quark, and therefore was not elementary, and could not carry the strong nuclear force.

Why can there be no Yukawa particle?

When Yukawa constructed the quantum function for his particle, he used a mathematical "real" function as a solution to the KG equation. It is important to note here that the notion of a "real" function was different from the admitted structure of quantum mechanics of massive particle. In fact, quantum mechanics uses what mathematicians call "complex functions." These functions are used as solutions of the quantum equations and describe the physical properties of massive particles.

There is a profound reason for using complex functions in describing massive particles. A massive particle can be at rest, so that a real function describing its state should be time independent. On the other hand, in quantum mechanics, the particle's variation with time is related to its energy. But for a particle which doesn't change in time, a quantum function of real numbers implies that the particle's energy (and mass) is zero. For this reason, a massive particle cannot be described using a real function, which means that the mathematical structure of Yukawa's particle is wrong from the start.

This error may seem fundamental, and people may find it hard to believe that a major physicist such as Yukawa could have gone wrong.

It turns out, however, that not only great physicists can and do make mistakes, but that this specific error is taught as a valid theory up to this very day and is featured in textbooks.[135]

A more precise and formal demonstration of this contradiction is evident in one of Comay's articles.[136]

It would be plausible to assume that the source of the error was the use of the calculation applicable to the photon, which is a massless particle, to calculate a massive particle. But this shortsightedness, discovered by Comay over 60 years later, proves that major errors may find their ways into basic textbooks and persist in the foundations of entire theories. Comay admits that he himself took part in this shortsightedness for decades, and only after discovering additional contradictions in Yukawa's theories, did he decide to

---

[135] H. Frauenfelder and E. M. Henley, *Subatomic Physics* (Prentice Hall, Englewood Cliffs, 1991). p.406
[136] E. Comay, *The Yukawa Lagrangian Density is Inconsistent with the Hamiltonian*, Apeiron **14**, No 1, 1 (2007).

thoroughly examine the theoretical foundations of Yukawa's particle.

The Klein-Gordon equation vs. experiments

In 1934 when Pauli and Weisskopf recycled the KG equation, very little was known about elementary particles, and in this respect, theoretical physicists were groping in the dark. Even Pauli admitted that his revived KG equation "had little to do with reality".

In the 1940s, several spinless particles were discovered, the most famous of which were a group of 3 particles called "pi mesons," or in short – "pions." The discovery of these particles was followed by a major rush of physicists toward the work of Pauli and Weisskopf.

In the 1960s, the existence of quarks was recognized and established. It turned out that these quarks actually compose the pions, among others, and that they were actually the building blocks of all the spinless particles discovered in strong interaction experiments. Evidently, a particle that is composed of other particles is not elementary. Hence, no spinless elementary particle is known today. Therefore, we are again at point zero, back in 1934 when Pauli complained that there were no known particles for which the KG equation could apply. On the other hand – the point-like massive elementary particle family grew substantially, and it turns out that they all have spin 1/2, as implied by Pauli's "enemy" – the Dirac equation of spin 1/2 particles.

The errors of the Klein-Gordon equation

The problematic nature of the KG equation is not limited to the real functions used by Yukawa. There are fundamental flaws in the

116

complex solutions of the equation as well.[137,138] Some of these flaws had already been mentioned by Dirac.[139]

Comay's article[140] mentions at least three additional contradictions to the KG equation. The most interesting of them shows how physicists, while being fully aware of the ensemble of physics' laws, fail to understand the deep connections between these laws. The point is this: the Lagrangian density and the Variational Principle yield a differential equation which constitutes a particle's equation of motion. Another differential equation can be obtained from a quantity called "Hamiltonian", which yields a first order time dependent differential equation.

A non-physicist reader should realize that Lagrangian and Hamiltonian are not simply the names of famous Armenian physicists. These are physical and mathematical notions, (named after the French mathematician Joseph Louis Lagrange and the Irish mathematician William Rowan Hamilton) which yield two differential equations describing the same particle. Both are derived from the same original equation – but each of them is obtained by a different mathematical development. Fortunately, as it turns out, in the case of Dirac equation, these two equations miraculously unite!

The KG equation wasn't as fortunate. Not only is it that the two equations deriving from the Lagrangian and the Hamiltonian do not

[137] E. Comay, *Further Difficulties with the Klein-Gordon Equation*, Apeiron **12**, No 1, 26 (2005).

[138] E. Comay, *The Significance of Density in the Structure of Quantum Theories*, Apeiron **14**, No 2, 50 (2007).

[139] P. A. M. Dirac, *Mathematical Foundations of Quantum Theory*, Ed. A.R. Marlow (Academic, New York, 1978). p. 3,4

[140] E. Comay, *Physical Consequences of Mathematical Principles*, Progress in Physics, **4** (2009). Chap.4.

unite, but they even end up contradicting each other. Therefore, the KG equation cannot be correct…

Another flaw of the KG equation specified in Comay's article (understandable only to physicists) is that no consistent Hilbert Space inner product can be constructed because the density function of a complex KG particle depends not only on its wave function, but also on the potential defined by an *external* source. This potential can vary with time and disrupt the definition of the Hilbert Space inner product.

One of the common features of natural laws is that when a theory is wrong – nature has many ways to show it.

Why there can be no Higgs

The Higgs Boson equation is a kind of an extension of the KG equation. Both equations describe a spinless particle. And the flaws characterizing the KG equation apply to the Higgs equation as well.

Therefore, the Higgs boson, predicted by a flawed mathematical instrument, does not exist.

The W and Z Bosons

The W and Z bosons are two spin-1 massive particles. The standard model assumes these two are elementary particles that carry the weak force. The standard model equations of the two particles suffer from problems similar to those of the Higgs Boson equations.[141]

---

[141] The issue can be understood by considering the Proca field, which is a classical massive field having a second order equation of motion. It is argued in the physics literature that this field describes a massive photon. It turns out that the kind of matter described by the Proca Lagrangian is fundamentally different than matter as it is known to science [E. Comay, Nuovo Cimento, **B113**, 733 (1998)]. This

This book aims to concentrate on electrodynamics and the strong interaction, and not get involved in the weak force issue. However, Comay recommends that the scientific community keep an open mind when interpreting the experimental results of W and Z bosons.

As an example of a different interpretation of W and Z bosons, the theoretical issues described above stem from the assumption that W and Z are elementary particles. To consider W and Z as composite particles would open the door to other interpretations, such as that W and Z particles are not elementary Bosons, but top-quark mesons. The fact that the top-quark is more massive than the W and Z particles could support this idea (we will see in the next chapters how the mass of a composite particle derives from its components). The absence of top-quark mesons from the PDG table of mesons could be an additional hint that W and Z are indeed top-quark mesons.

By the way, assuming that the W, Z Bosons are mesons containing top quark, it is possible to find more mesons containing top quark in higher energy levels. It is therefore possible that in the future we will find particles with higher mass than the W, Z Bosons and shorter half-life period.

Comay believes that the last word has not yet been said regarding these questions, and that it is worthwhile to leave them open for further consideration.

---

discrepancy casts doubt on its validity, and may constitutes one theoretical reason behind the failure of finding such matter.

UNIT 3: ADDITIONAL MASS IN THE PROTON

Chapter 12: Something Else inside the Proton

During the 1960s, physicists generally came to agree that protons and neutrons contain three quarks. Several years later experimental findings showed that only half of the proton mass is carried by the quarks.

In this chapter we'll see evidence for additional massive objects inside the proton. As modern particle accelerators reach increasingly higher energy levels, this becomes more and more apparent.

An inspiring quantum mechanical experiment

In 1913, James Franck and Gustav Hertz conducted one of the first experiments showing that the atom's energy does not take arbitrary values, but rather specific energy levels. The results of the Franck-Hertz experiment supported Bohr's atomic model. Bohr's model was later abandoned, but the results of their experiment gave a major boost to the development of the nascent quantum mechanics.

Franck and Hertz fired electrons into a tube containing argon gas or mercury vapors, and measured the value of the electric current at the other end of the tube. The electrons were accelerated by means of an electric field.

Franck and Hertz were able to control the electrons' velocity by varying the electric field. At first, for lower values of electric field, the current increased consistently with the increase of electron speed. But when the electrons' velocity reached a certain threshold, the current at the other end of the tube suddenly dropped. When they further increased the electric field and the acceleration of the electrons, the current at the end of the tube rose again, to a certain threshold, and then, when electrons' speed crossed this new threshold, the current level dropped again.

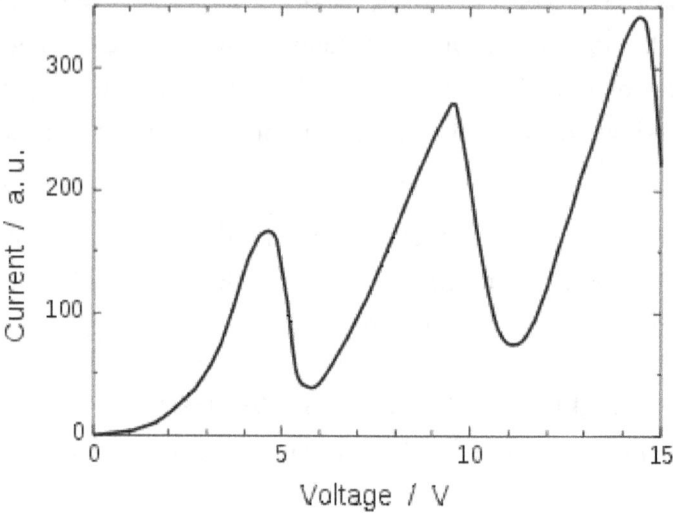

Figure 27. The current in the Franck–Hertz experiment

What were the experiment's conclusions? An atom's electrons can only have well defined specific energy levels. When an incident electron hits an atomic electron, it loses a part of its kinetic energy by transferring it to the atomic electron. When the energy absorbed by the atomic electron is high enough, it reaches the threshold which raises the atom to a higher energy level. In such a case we say that the incident electron excites the atom.

In the Franck-Hertz experiment, when the incident electrons were moving slowly and their energy was too low to excite the atoms, the

124

electrons simply traveled across the atoms and reached the end of the tube with no energy loss. But when the incident electrons were accelerated and reached a higher velocity (corresponding to higher kinetic energy), their energy was high enough to excite the atom by transferring energy to the electrons inside the atom. The current at the end of the tube dropped because after transferring energy, the electrons had less energy by the time they arrived at the end of the tube. This explains the first drop in current intensity. The second drop in current intensity corresponds to the loss of electrons' kinetic energy when exciting two atoms, and so forth.

Analogous results are regularly observed in today's particle accelerators as well, but at energy levels much higher than those used a century ago. An important aspect of the analysis of these experimental results is based on what physicists call "the cross section curve," which expresses particle collision rates and is related to the probability of observing a specific event as a result of the collision. It turns out that the cross section depends not only on the type of particles involved in the collision but on the process energy as well.

The study of the proton's structure

In the Franck-Hertz experiment, incident electrons colliding with atoms interact with atomic electrons. Exploring the proton in a similar way requires that the incident particle energy be significantly higher, for two closely-related reasons.

One reason is that the energy levels of the quarks inside the proton are much higher.

The second reason is that in quantum theory, a particle's location is not point-like, as in classical physics, but is described by a "wave".

A slowly-moving particle has a relatively large wavelength. Particles can only interact with objects of dimensions comparable to their wavelength. A smaller wavelength of the incident particle enables an examination of smaller constituents at the target proton.

Thus, when a particle's wavelength is of the order of the proton's size, it collides with the whole proton and not with an isolated quark inside the proton. Therefore, in order to study quarks, the incident particle's wavelength needs to be significantly smaller than the size of the proton. Since the particle's wavelength is shorter when the particle's energy is higher, the incident particle must move at a very high energy in order to interact with a single quark inside the proton.

The first accelerator capable of reaching such energy levels was built at the Stanford Linear Accelerator Center (SLAC) where electron beams were accelerated over a straight line of two miles length. During the last few decades, scientists succeeded in accelerating many different particles, such as electrons, positrons, muons and protons, to very high energies which significantly shorten the particle's wavelength, allowing these particles to interact with a single quark within the proton.

What happens when a proton is bombarded?

When a high-energy particle beam hits a proton, two types of interactions between the incident particle and the proton can occur.

One type of collision is called "elastic collision", where kinetic energy and momentum are exchanged between the particles. The products of an elastic collision are the same particles that existed before the collision.

The second type of collision, called "inelastic collision", is characterized by the creation of new particles, such as mesons, composed of a quark-antiquark pair. This phenomenon was explained in the chapter dealing with quark confinement.

Scientists observe that when they raise the energy levels of electron-proton collisions, nearly all the collisions are "inelastic". In fact, at the high energy range, as energy increases, the relative rate of elastic collisions decreases and reaches a tiny fraction of about 1:1000.

In an electron-proton collision, the probability of inelastic collision also decreases with the increase of energy, but at a much lower rate. This pattern was consistently observed until experiments in the late 1990s reached unprecedented energy levels.

Late 1990s, DESY laboratories, Germany

The results of experiments conducted at the DESY Labs in Germany were published in 1997. These experiments consisted of bombarding protons with beams of electrons or positrons, at higher energy than ever before. The experiment was performed from 1994 to 1996 by two independent groups of researchers, at two different locations of the device. The results obtained from both experiments where similar, and surprising.

Physicists expected to see a decrease in the cross-section curve for higher energies, because of a fundamental rule established about a century ago, stating that the probability of a collision between an electron (or a similar particle) and a proton decreases as the incident particle's energy increases.

However, the results here showed an unexpected relative increase in the number of collisions as energy increased. The discovery was followed by the publication of two papers expressing astonishment with regard to these findings. [142,143]

Frank Sciulli, professor of physics at Columbia, who drafted one of the papers reporting the observations, said: "If the results are not a statistical fluke, new physics has been observed. One possibility is that our understanding of what's inside the proton is somehow wrong."[144]

What are the possible explanations for this increase in interactions when the beam's energy goes up?

One possibility would be that in analogy to the Franck-Hertz experiment, there are other entities inside the proton capable of absorbing this energy only above a certain threshold. This explanation obviously contradicts QCD.

The groups conducting the experiments concluded, quite reasonably, that their findings might be accidental, because not enough data were collected to be clearly statistically significant.

Early 2000s, Fermilab, Illinois

In an experiment, conducted some 10 years ago in the Tevatron Collider at Fermilab in Illinois, USA, a proton beam collided with another proton (or anti-proton) beam, at energy levels higher than those reached at DESY, with clear cut results. The proton-proton

---

[142] C. Adloff *et al.*, *Observation of Events at Very High Q2 in ep Collisions at. HERA* , Z. Phys C74, 191 (1997)

[143] J. Breitweg *et al.*, *Comparison of ZEUS Data with Standard Model Predictions for ep -> eX Scattering at High x and Q2*, Z. Phys C74, 207 (1997)

[144] Columbia University News, 1997. www.columbia.edu/cu/pr/97/ 19058.html

cross-section curve does indeed stop decreasing and begins to move up when the beam's energy goes beyond a certain threshold.

Furthermore, the rate of elastic collisions increased as well. In other words, a change of tendency was observed both in the total and the elastic cross sections, and their curves changed directions and started increasing.[145]

---

[145] C. Amsler *et al.* (Particle Data Group) Physics Letters B667, 1 (2008). Also see p.12 in pdg.lbl.gov/2009/reviews/rpp2009-rev-cross-section-plots.pdf

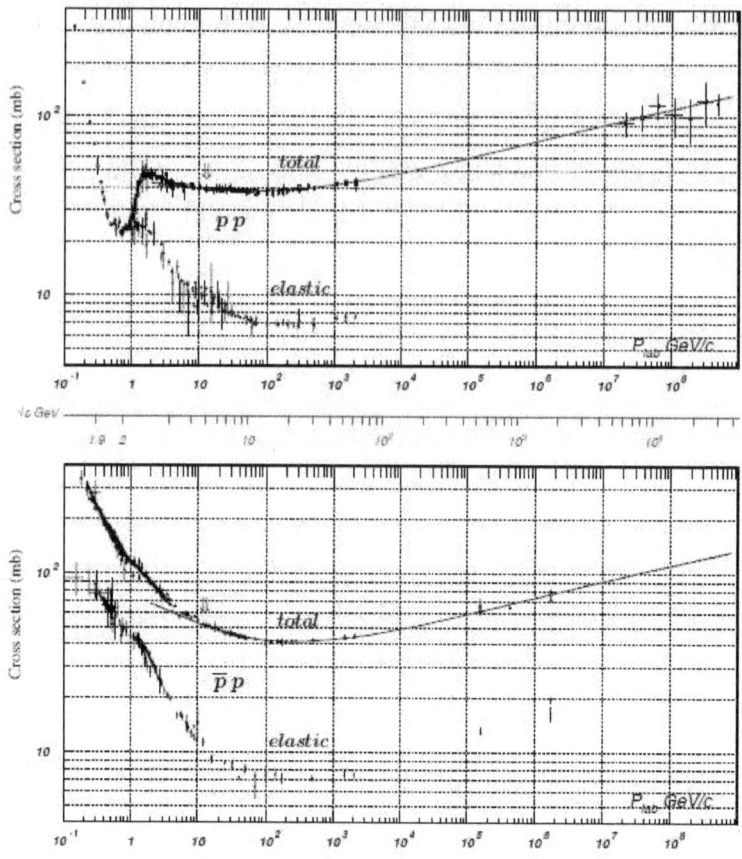

Figure 28. Total and elastic cross sections for *pp* and *pp̄* collisions as a function of laboratory beam momentum and total center-of-mass energy.

This finding has been known for several years now and is supported by earlier results. The only plausible explanation for this is the existence of massive particles inside the proton, which get "excited" at this high energy level and produce this tendency change in the curve. In analogy to the Franck-Hertz experiment, these massive particles can only occupy specific energy levels, and react to the collision only when the energy of the incident particle is high enough to excite them and move them to a higher energy level.

130

An essential element of QCD states that protons (and all other baryons) consist of the three valence quarks, additional quark-antiquark pairs and gluons. In this theory there is no room for another proton component that enters the interaction only when collision energy is higher than a certain threshold. Therefore, there is no explanation for this fundamental effect within the framework of QCD.

Scientists currently ignore this experimental result. Very few scientists make an issue of the fact that QCD does not provide an explanation for this phenomenon.[146,147]

Comay's model naturally accounts for this finding.

Inside the proton there is a core with closed inner quark shells. Whereas moderately high-energy experiments could only detect the quarks located at the external shells, higher energy particle beams were able to excite the quarks in inner quark shells, in analogy to a multi-electron atom.

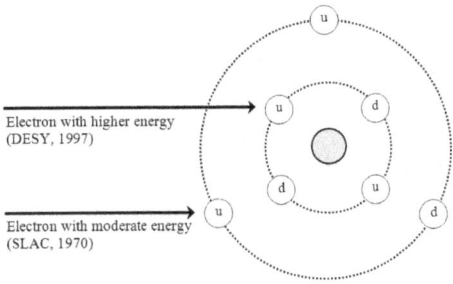

Figure 29. Explaining DESY experiment by Comay's model

---

[146] A. A. Arkhipov, *On global structure of hadronic total cross sections*, 1999. arxiv.org/PS_cache/hep-ph/pdf/9911/ 9911533v2.pdf

[147] E. Comay, *On the significance of the upcoming large hadron collider proton-proton cross section data,* Prog. in Phys. 2, **56** (2010).

More about elastic scattering

The consensus today is that high-energy scattering of electrons on protons indicates that the interaction actually takes place between the electron and a single quark inside the proton.

According to the rule mentioned above, in such a case, the relative rate of elastic collision should decrease with the increase of the energy, and become negligible for very high energies. This would mean that if a quark is heavily hit by an electron, the proton almost always undergoes an inelastic collision.

But it turns out that in high-energy proton-proton collisions, the rate of elastic scattering events is about 15% and for collision energy used before 2011, it does not decrease as energy increases. These results lead to the following conclusions:

a. The proton contains a rigid component which is not a valence quark, and which can resist high-energy collisions and maintain the proton's integrity.

b. This component is electrically neutral and is therefore not revealed in electron-proton scattering.

c. The proton's rigid component participates in the interaction only if the collision energy is high enough. The new channel opened at higher energy increases the number of events and the cross-section curve starts rising.

These conclusions validate the existence of a core inside the proton, containing closed quark shells and offer an explanation for the fact that the proportion of elastic collisions in proton-proton scattering is far from being negligible.

The near future, CERN

The proton beam of the LHC particle collider at CERN reaches even higher energy levels, which will further reveal the behavior of the cross section curve as energy goes up.

According to Comay[148], if the number of inner shells is small, or if the energy needed to excite quarks in even deeper shells is higher than the experimental energy, then the elastic cross section will start decreasing again.

Alternatively, if the energy levels obtained during the upcoming CERN experiment are able to excite even inner quark shells, it will bring out the next zone on the curve in which the elastic scattering cross section goes up.

A prediction

The increase of the cross section graph clearly suggests that the proton has a complex massive core.

Comay suggests another experiment, measuring the cross section of energetic pion-pion collisions.

Comay predicts that unlike the proton-proton cross-section curve, the cross section curve of pion-pion collisions will decrease continuously. This is simply because mesons do not contain inner shells.

---

[148] *Ibid.*

Chapter 13: The Mass of the Valence Quarks

As mentioned earlier, an experiment conducted at the Stanford Linear Accelerator Center during the early 1970s demonstrated that the valence quarks (together with the quark-antiquark pairs) account for about one half of the nucleon's mass. We shall see in this chapter and the following one how this experimental result fits Comay's model and how an analogous experiment is expected to yield decisive results.

According to QCD, baryons contain three quarks, and no other massive particles. Comay's model assumes that baryons have a massive core and that the three observable quarks occupy an outer energy shell. Beside the three valence quarks, there are internal closed shells of quarks.

With regard to mesons, both models agree that they are composed of a quark-antiquark pair and contain no other massive particles. It is further agreed that according to Field Theory, there is a probability of finding additional quark-antiquark pairs inside baryons and mesons.

In Comay's model, the dynamical properties of quarks are analogous to the electromagnetic energy relations that hold for electrons. The following energy notions are useful for describing energies of bound systems of particles.

Useful energy notions

Electrons within the atom are bound to the nucleus, and the hadrons are bound states of quarks. In physics, four different notions relate to the energy of these states:

A bound state can only occur if "attraction forces" apply between the system's components. The energy derived from these forces is called "potential energy". The potential energy of a bound system has a negative value.

In physics it is known that a bound particle cannot be at rest. For example, Earth cannot be at rest within the sun's gravitational field. Quantum theory considerations lead to the same conclusion for electrons in atoms and for quarks in hadrons. Bound particles therefore have energy as a consequence of their motion. This energy is called "kinetic energy". Kinetic energy is always positive.

The sum of the (positive) kinetic energy and the (negative) potential energy, indicates the bond strength of the system's components. This is called "Binding Energy".

The "total energy" (or total mass) of a bound system is the sum of the self mass of its constituents minus the binding energy. It is a positive quantity.

Mass of bound particles

When we look at atoms, the mass of the whole atom is almost exactly equal to the sum of the masses of the atom's components.

According to special relativity mass and energy are equivalent. Therefore, the mass of an atom should be the sum of the masses of its nucleus and electrons minus the binding energy. And indeed, the mass of an atom is a bit smaller than the sum of the masses of its components.

In hadrons, the situation is the same in principle, but the binding energy plays a critical role in the hadronic mass. Although free u or d quarks were not found experimentally, we can assume that such particles have a certain mass if they are free and at rest.

Let's take the positively charged, pi meson (π+) as an example. A π+ contains u and $\bar{d}$ quarks, and its total mass is 139.6 MeV. According to the principles above, this mass is the mass of u quark, plus the mass of d quark, minus the binding energy. Since the quarks are bound so strongly to each other, the binding energy is very high, and therefore the mass of the hypothetically free u or d quarks is probably thousands MeV.

Quark mass in pi mesons and protons

We will start with the natural assumption that u and d quarks have nearly similar masses. Indeed, if we look at mesons and baryons we can see that whenever we replace all u quarks by d quarks and vice versa, in such a particle the total mass remains almost the same. Therefore, this assumption seems quite reasonable. As is well known (to physicists), this assumption relies on the usefulness of the isospin symmetry.

Let's first look at the pi meson and see what we can deduce about its quarks' mass. The total pi meson mass is 139 MeV. We can assume, according to field theory, that as in the proton, a pi meson contains additional pairs of quark and antiquark. Therefore, it is reasonable to assume that in the case of a pi meson, the mass of each quark in bound state is about 50 MeV.

For the proton we know that there is nearly one half of additional quark-antiquark pair. Therefore, the number of quarks plus antiquarks in the external shell is about 4. The proton's mass is 938

MeV and quarks make up half of it, about 470 MeV. This means that in the case of the proton, the mass of a quark in bound state is about 120 MeV.

Why is the mass of a bound quark in the proton nearly twice as large as its mass in a pi meson?

Potential energy of valence quarks

We should remember that in a bound system, the mass of a quark is its rest mass minus its binding energy. The binding energy is the difference between the absolute value of the negative potential energy and the positive kinetic energy of the quark.

Let us first look at the potential energy of a quark inside a pi meson. According to Comay's model, the quark is attracted to the antiquark because they have opposite strong charges. In an appropriate unit system, the unit of strong charge of the antiquark is +1 and that of the quark is -1.

In the case of the proton, the quark is attracted to the proton because the total of strong charges of the proton's core is +3. It attracts three quarks, each of which has a negative unit of strong charge. This means that the potential energy of the quark inside the pi meson is lower than that of the proton.

Similar effects are known in atoms where electromagnetic forces define the system's characteristics. For example, in the case of atoms, each of the Helium's two electrons is bound more strongly to the helium nucleus than the bond of an electron in the hydrogen atom.

In analogy, we can assume that the quark is bound to the proton a bit more strongly than to the pi meson.

The kinetic energy of the valence quarks

In quantum mechanics, the kinetic energy of a bound particle can be estimated according to the spatial volume in which it is contained. It is well known that the kinetic energy of a particle bound within a smaller volume, is higher.[149] For example, the kinetic energy of an electron in the helium atom is higher than the kinetic energy of an electron in the hydrogen atom, because the volume of the helium atom is smaller, and therefore the electron is more "restricted" than in the hydrogen atom.

But if we examine heavier atoms with several electron shells, we see that the electrons in the external shells have larger kinetic energy even though their volume seems to be larger. This is explained by the fact that, due to the Pauli exclusion principle, the electrons in the external shell are restricted by the electrons of inner shells, and their kinetic energy is therefore higher than that of a single electron enclosed inside the entire volume of the same atom.

According to Comay's model there are inner u,d quark shells in baryons. Therefore, similar arguments can apply to hadrons. The pi meson would be analogous to the hydrogen atom and to the positronium (because each of these systems consists of two interacting particles) whereas the proton is analogous to an atom with several electronic shells.

Moreover, the size of the proton is not much larger than that of the pi meson. Thus, due to the proton's inner closed shells of quarks, it is reasonable to assume that here too the kinetic energy of the valence quarks in the proton is much higher than their kinetic energy in the pi meson.

---

[149] This is derived from the uncertainty principle.

This argument completes the explanation of the experimental finding obtained from Stanford Linear Accelerator Center, where it was shown that the proton's quarks carry about one half of the proton's mass. As mentioned above, in the proton, the single quark energy is much greater than the corresponding pi meson value. It also provides an explanation for the relatively small mass of pi mesons.

Chapter 14: A Decisive Experiment

Here we suggest an experiment promising to contradict at least one of the strong interactions theories: QCD and Comay's model.

The theories

According to QCD, baryons contain three quarks, and no other kind of massive particles. Baryons also contain gluons, which glue the quarks together and make the proton stable (see Figure 30).

Comay's model, on the other hand, assumes that baryons have a massive core and that the three observable quarks occupy an outer energy shell. Beside the three valence quarks, there are inner closed shells of quarks. The system is analogous to a multi-electron atom (see Figure 31).

Both theories agree that based on field theory there is a probability of finding additional quark-antiquark pairs inside hadrons.

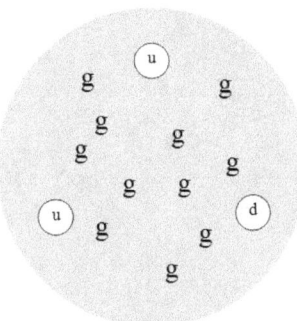
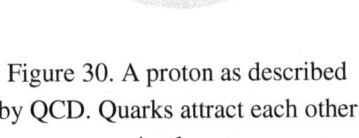

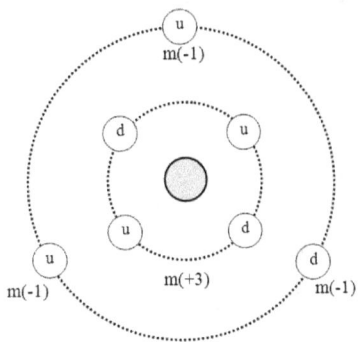

Figure 30. A proton as described by QCD. Quarks attract each other via gluons

Figure 31. A proton according to Comay's model. Quarks are attracted to the proton core because they have opposite sign of strong charge (in analogy to a non-ionized atom). The 4 inner *u,d* quarks represent a minimal inner quark closed shells

With regard to mesons, both models agree that they are composed of a quark and antiquark and contain no other kind of massive particles. (see Figure 32, Figure 33). Here too, according to Field Theory, there is a probability of finding additional quark-antiquark pairs inside the mesons.

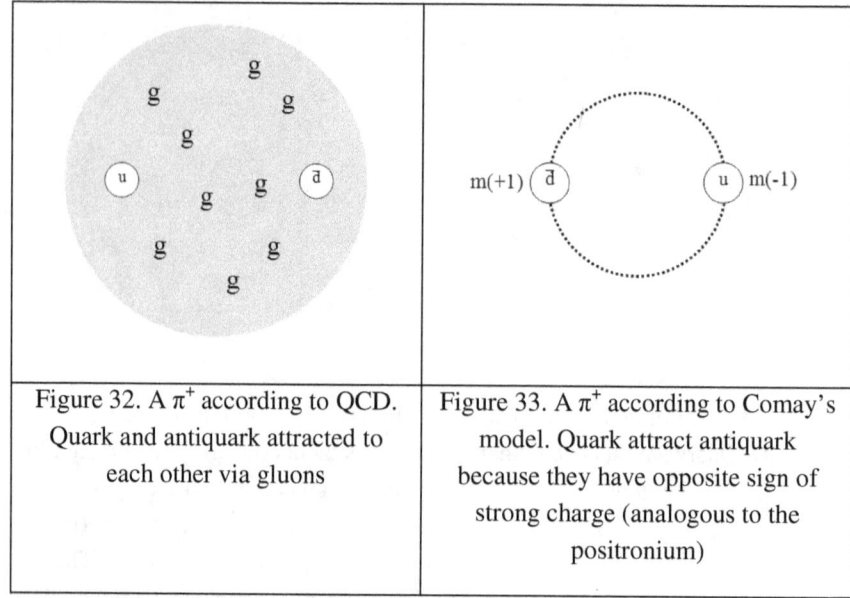

| Figure 32. A π^+ according to QCD. Quark and antiquark attracted to each other via gluons | Figure 33. A π^+ according to Comay's model. Quark attract antiquark because they have opposite sign of strong charge (analogous to the positronium) |

The experiment

An experiment conducted at Stanford Linear Accelerator Center (SLAC) during the 1970s, where an electron beam hit a proton target, showed that the quarks and the quark-antiquark pairs carry only about one half of the proton's mass. According to QCD, the gluons carry the other half of the proton's mass.[150]

If QCD is correct, then in the case of the pi meson, which consists of a quark and an antiquark glued by gluons, the gluons should carry a non-negligible part of the mass of the pi meson as well.

On the other hand, in Comay's model, there are no gluons and mesons have no core. Therefore, in the case of meson, the entire mass should be found in the quark-antiquark components.

---

[150] The gluons as described by QCD are massless particles that carry roughly half of the proton mass.

The decisive experiment would thus be similar to the 1970s SLAC experiment, but this time an energetic electron beam would collide with a pion beam. Analysis of the result of the deep inelastic collision would yield one of the following possibilities:

- All the mass of the pi meson can be attributed to the quark-antiquark components (QCD is refuted).

- The gluons carry roughly one half of the pi meson mass. (Comay's model is refuted).

- The pi meson mass is incompatible with both of the above (both theories are refuted)

Today, it is possible to produce energetic pi meson beams and analyze the results of a deep inelastic collision of a pion beam and an electron beam. Only are left to be done: carry out the experiment, and bravely make the necessary conclusions based on its results.

Chapter 15: Baryon Radius too Large

One of the interesting physical dimensions is the estimate of the radius of a quantum particle composed of bound sub-particles. In quantum mechanics the boundary of a composite particle is not well defined. Therefore, one cannot find a sharp definition for the radius of such a particle. However, there are good methods for finding average density that yield an estimate for the radius. The particle's radius often provides us with hints to understanding the particles' structure and the interactions between its components.

In order to understand how this information can be of use, let us take a look at the structures of well-known particles, considered as quantum systems of electric charges, and compare them to what we know about hadrons, which are particles composed of quarks.

Radii of hydrogen, helium and lithium atoms

Let consider a few bound quantum states, ruled by the laws of electromagnetism. The simplest case is that of the hydrogen atom, composed of a proton and an electron. Another case is the helium atom, which has two electrons and a nucleus containing a double charge (2 protons + 2 neutrons or 2 protons + 1 neutron).

Table 4. Radii of atoms.

Radii of atoms[151]	
Hydrogen (H)	0.53 angstrom[152]
Helium (He)	0.30 angstrom
Lithium (Li)	1.67 angstrom

Measurements showed that the radius of the helium atom is smaller than that of the hydrogen atom.

This is because the helium nucleus has two protons, applying a stronger attraction force on the electrons and bringing them closer together, thus condensing the helium atom such that its radius is smaller than that of the hydrogen atom.

One may therefore conclude that the higher the number of protons in the nucleus, the stronger the interaction, and the smaller the radius.

But the radius of the lithium atom, containing three protons and three electrons, is larger than that of the hydrogen atom. Why?

The lithium's electrons occupy two shells, whereas the electrons of the hydrogen and the helium atoms occupy a single shell. Indeed, in the case of one or two electron atoms, the Pauli exclusion principle allows electrons to be in the lowest shell. On the other hand, in the three- electron lithium atom, one electron occupies a higher shell and the atomic radius increases.

---

[151] S. Fraga, K. M. S. Saxena and J. Karwowski, *Handbook of Atomic Data* (Elsevier, Amsterdam, 1976). p. 467
[152] 10,000,000,000 angstroms is one meter. Named after the Swedish physicist Anders Jonas Ångström (1814–1874).

Calculations based on quantum theory confirm these results. It therefore seems that the atom's radius is determined mainly by the geometrical size of its external electronic shell.

Radii of the pi meson and the proton

The proton is characterized by a three-quark combination. The pi meson, (or the pion for short), is composed of a quark-antiquark pair of the same flavors as those composing the proton.

Experimental findings show that the pion's radius is slightly smaller than that of the proton.

Why? Comay's model claims that the proton has a core carrying three strong charges that attract three quarks and that this core contains inner closed shells of quarks. The existence of inner shells could explain the proton's larger radius, in analogy to the lithium atom, which has a larger radius because its electrons fill two shells, and rather than one single shell.

As explained above, Comay's model claims that all baryons have an inner core containing closed shells of quarks, in analogy to a multi-electron atom.

Due to the extremely short life span of most hadrons, the radii of only a few hadrons have been measured experimentally, but the results of baryon and meson measurements are fully consistent with Comay's model. As in the case of the comparison between the proton and pion radii, the comparison between the Σ^- baryon and the K^- meson (both containing an s-quark) shows that the K-meson's radius is smaller. The mean charge radius of K^+ and K^- is 0.560 fm and that of Σ^- is 0.78 fm.[153] This is consistent with the assumption that the baryon contains inner shells.

One additional comment about the radius of hadrons

We saw above that a particle's radius is determined by the number of its sub-particles, and in particular by the distribution of these sub-particles in shells.

But it also depends on the sub-particles' mass. We already know from observing the electromagnetic force that a particle composed of a proton and a muon (the muon is heavier than the electron) has a smaller radius than that composed of a proton and an electron. This effect is consistent with the fundamental properties of an electromagnetically bound system as described by quantum theory. It means that when two oppositely charged particles are bound together, the composite particle's radius will be smaller if its components' mass is larger.

Comay's model draws parallels between the properties of the electromagnetic force and the strong interaction. Therefore, mesons and baryons under strong interactions should manifest similar phenomena to their properties under electromagnetic forces.

And indeed, meson radius measurements show that the K^+ meson radius, for example, is smaller than that of the π^+ meson, because the s-quark is heavier than the d-quark, in analogy to the fact that the Σ^- baryon's radius is smaller than the proton's radius.

---

[153] K. Nakamura *et al.* (Particle Data Group), J. Phys. G **37**, 075021 (2010)

Chapter 16: Simple Mass Calculation

Comay's model compares quarks in the baryon to electrons in the atomic nucleus with regard to the forces acting on them. It also compares mesons to positroniums, which is the electron-positron bound state.

The model further assumes that every quark carries a unit of negative strong charge, and the baryonic core carries 3 units of positive strong charge.

Calculating the actual quantum states under strong interaction is an extremely difficult task and goes far beyond the scope of this chapter. We will therefore use simple arithmetic to examine the parallels drawn by Comay's model between the different systems: the atom versus the baryon, and the positronium versus the meson.

With regard to systems with electrons, we know that the binding energy of an electron and a proton (the hydrogen atom) is less than half of the binding energy of two electrons bound to two protons (the helium atom). This means that the electron's bond to the helium atom is stronger than its bond to the hydrogen atom. We can see that by examining the energy required to ionize one electron from an atom. This energy is called the "first ionization energy."[154]

---

[154] www.orschemistry.com/PVExp5-2c.pdf

Table 5. 1st ionization energy.

Symbol	1st ionization energy (eV)
H	13.6
He	24.6
Li	5.4

We can see that in spite of the fact that the nucleus of the Li atom has three protons and it attracts electrons stronger than the nuclei of the He and H atoms, the energy required for ionizing the first electron of the Li atom is much less than either that of the He or the H atom. This is due to the Pauli exclusion principle. The third electron of the Li atom has a larger radius and its kinetic energy is higher than that which is related to its volume. Thus, its binding energy is smaller and less energy is required for ionizing it.

So far, the masses of many baryons and mesons have been successfully measured. In this chapter we'll show how Comay's model fits these findings.

What determines a particle's mass

The total mass of the nuclear building blocks, the neutrons and the protons, usually provides a good approximation of the nuclear mass.

By definition, the nuclear mass is equal to the sum of its constituent masses minus a mass equivalent to the binding energy between them. This is because it takes energy to separate 2 bound particles.

Let us look, for example, at the helium-4 nucleus, which is composed of two protons and two neutrons. Its mass is 3727 MeV and the mass of its constituents when they are not bound, two free protons and two free neutrons, is nearly 1% larger: about 3756 MeV (2x938.3+2x939.6).

The binding energy of protons and neutrons inside atomic nuclei is less than 1% of their global mass, and one can therefore consider that the mass of protons and neutrons in the nucleus is very close to that of free protons and neutrons.

The situation in hadrons is much more complex.

The magnitude of the strong interaction is much stronger than the nuclear force, and therefore the binding energy in hadrons is much higher.

Theoretically, high binding energy means that separating the quarks from each other would require a huge amount of energy. Therefore, the mass of a hypothetical free quark would include the mass equivalence of the binding energy of the bound quark. Therefore, if the quark could exist outside the hadron, its mass should be much greater.

This is very different from what we know about electromagnetic and nuclear forces. The binding energy of quarks inside the hadrons is a major component of the composite's particle mass, on top of the masses of the quarks composing it, which means that the difference in mass between a bound and a free quark is very significant.

If we overcome our reluctance to perform calculations, and try to explore the mass of several types of hadrons, we may discover some interesting things. The simple calculations presented here confirm that the mass of baryons and that of mesons behave according to well-known properties of electromagnetically bound systems, which is perfectly consistent with Comay's model.

Baryons compared to mesons

Data tables specifying baryons' and mesons' masses can be found on the PDG web site[155] or in Wikipedia.[156] Listed below are the masses of baryons and mesons which are important for our discussion.

Table 6. Masses of baryons and mesons.

Baryons			Mesons		
Name	Quarks	Mass	Name	Quarks	Mass
Proton	uud	938.3	π^+	$u\bar{d}$	139.6
Neutron	udd	939.6			
Σ^+	uus	1189	K^+	$u\bar{s}$	494
Charmed Σ^{++}	uuc	2454	D^0	$u\bar{c}$	1865
Bottom Σ^+	uub	5808	B^+	$u\bar{b}$	5279
Σ^-	dds	1197	K^0	$d\bar{s}$	498
Charmed Σ	ddc	2454	D^-	$d\bar{c}$	1870
Bottom Σ^-	ddb	5815	B^0	$d\bar{b}$	5279

* mass unit is MeV

[155] K. Nakamura *et al.* (Particle Data Group), J. Phys. G 37, 075021 (2010) pdg.lbl.gov/
[156] en.wikipedia.org/wiki/List_of_baryons
en.wikipedia.org/wiki/List_of_mesons

Let's try to compare between the masses of baryons and mesons. We will consider the lightest particle composed by a set of quarks, which is the most stable state of each quark combination.

We start by comparing a baryon – the proton, to a meson – the π^+ particle. The proton has a mass of 938 MeV and the π^+ has a mass of 139 MeV. The proton contains 3 quarks: *uud*, and π^+ is composed of *ud̄* quarks. Everyone agrees that an antiparticle's mass is identical to the particle's mass, and only their charges are opposite to each other.

According to Comay's strong interaction model, the proton contains a complex core plus three quarks on the external shell. It is therefore not surprising to find that the proton is heavier than the quark-antiquark state of the π^+. As we demonstrated in the chapter about the mass of the valence quarks, the difference in mass between the proton and the pion exists for two different reasons: the baryonic core is nearly 470 MeV whereas the pions have no core, and the kinetic energy of the proton's valence quarks is higher because of the *u* and *d* quarks in the inner quark shells. Indeed, as in the case of the Li atom, the *u, d* valence quarks reside in an external shell. Hence, their binding energy is less and the proton's mass is greater.

Binding energy of baryons and mesons

The difference between the energy of mesons and baryons composed of similar quarks is so great, that people who do not accept Comay's model might think that it is explained by binding energy, meaning that mesons have significantly higher binding energy than baryons.

We will see now that for *s, c, b* quarks the opposite is correct: the binding energy of baryons is higher than that of mesons. This is in line with Comay's model in which the proton's core consists of closed *u, d* quark shells and no other quark flavor. Hence, in the case of baryons, the Pauli exclusion principle does not hold for quarks having other flavors. We will see now that contrary to the case of the *u, d* quarks, in other quark flavors the binding energy of baryons is higher than that of mesons, and replacing a quark with a heavier quark increases the binding energy of the baryon more than that of the corresponding meson. This rule is in accordance with quantum mechanics of electromagnetic systems.

Let's consider another baryon and meson pair. The Σ^+ baryon is composed of the quarks *uus* and its mass is 1,189 MeV. The K+ meson, composed of $u\bar{s}$ quarks, has a mass of 494 MeV. What does this mean? It turns out that each time an *s*-quark replaces a *u* or *d* quark, the resulting particle is heavier. Therefore it is admitted that the *s*-quark is heavier than the *u* and *d* quarks.

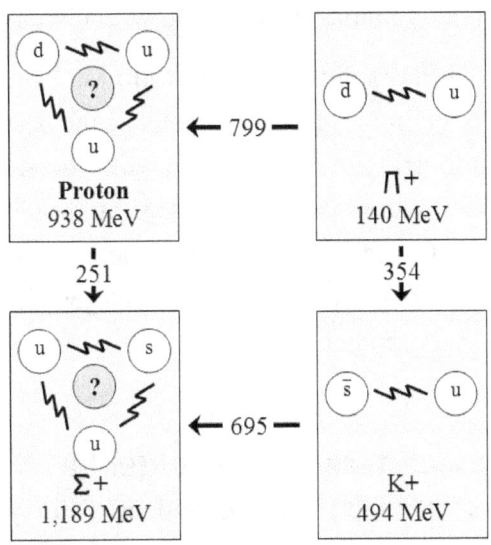

Figure 34. Mass value and mass difference of two baryons and two mesons

Figure 34 that the mass of a baryon in which an *s*-quark takes the place of a *d*-quark increases by 251 Mev (from 938 to 1189). In the

153

case of the meson, when an *s*-quark replaces a *d* quark, the meson's mass increases by 354 MeV (from 140 to 494).

This means that replacing the *d*-quark with an *s*-quark increases the baryon's mass relatively less than it increases the corresponding meson mass. This suggests that for an *s* quark, the binding energy of the meson is lower than the binding energy of the baryon.

And indeed, just as the binding energy of electrons inside the helium atom is higher than that in the hydrogen atom, the *s*-quark inside the baryon is bound more strongly to the inner core carrying three units of strong charge, than in the meson, where it is attracted only to one quark carrying only one unit of strong charge. And since a stronger attraction force increases the binding energy and decreases the particle's mass, this result as well is consistent with the .general laws of electromagnetically bound systems and with Comay's model.

This discussion can be summarized as follows. Beside the valence quarks, baryons contain closed quark shells of the *u, d* flavor. Due to the Pauli exclusion principle, the proton (and the neutron) valence quarks are pushed to outer regions and their binding energy reduces. This effect does not hold for an *s* quark. Hence, the three units of strong charge of the baryonic core make a larger binding energy of an *s* quark to a baryon than mesonic bond of an $\bar{s}$ to a single *u* quark.

Is relative mass difference coincidental?

The relative mass difference is not coincidental. Comparison to other baryon-meson pairs containing heavier quarks, such as Σ^{++}_c composed of the *uuc* quarks, and D^0 composed of $u\bar{c}$ quarks, or Σ^+_b composed of *uub* quarks and B^+ composed of $u\bar{b}$ quarks, reveals exactly the same phenomenon.

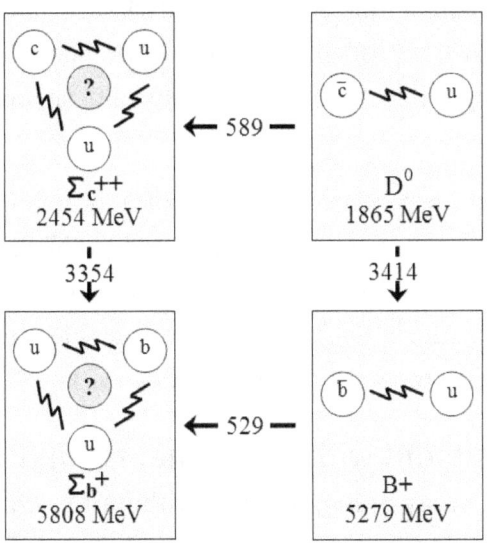

Figure 35. Another example of mass value and mass difference of two baryons and two mesons

Thus, in a baryon, the binding energy of a heavier quark minus the binding energy of a lighter quark is greater than the corresponding difference in a meson.

Any comparison between a baryon and its corresponding meson shows that, with the exception of the u,d quarks[157], the same quark is more intensely bound to the baryon than to the corresponding meson. That is, here too, binding energy is coherent with Comay's model.

In this chapter we used simple considerations to try to understand a complex problem. Other parameters play non-negligible roles in determining a particle's mass, such as the probabilities of

---

[157] The exception of the u,d quarks is related to the inner closed shells of these quarks and to the effect of the Pauli exclusion principle.

configurations containing additional quark-antiquark pairs. (It has been shown that the probability of the existence of such additional pairs is non-negligible in baryons. They are supposed to exist in mesons as well.)

Above all, a high-precision calculation of a system characterized by the strong interaction is a very complicated problem. Here we made a qualitative calculation taking into account higher or lower interaction intensities. Such qualitative considerations provide an interesting insight into the real situation.

Is QCD consistent with experimental findings?

QCD supporters may argue, and not without reason, that their laws relative to the strong interaction are very different than the electrodynamic laws describing the electrons in atoms. But the consistent gap between baryons and mesons, and the way masses and radii correspond both to what we generally know about systems bound by electromagnetic forces and to Comay's model, should ring yet another bell with regard to the validity of QCD.

Chapter 17: Charge Radius of Σ^+ Baryon

You can see that Comay's model and QCD describe baryons very differently. This book contains many examples of known experimental results which are in line with Comay's model and are inconsistent with QCD.

The differences between these theories suggest that it should not be too difficult to design new experiments for which QCD and Comay's model would predict different results. One such experiment is the measurement of the "charge radius" of the Σ^+ baryon. We will explain later how charge radius is calculated.

The Σ^+ baryon is characterized by *uus* valence quarks, and has several similarities to the particle Σ^- which is characterized by *dds* valence quarks. The charge radius of Σ^+ was not measured experimentally until November 2010. The charge radius of Σ^- was measured at the beginning of the 2000s.[158] According to a QCD prediction, the charge radius of Σ^+ baryon should be within the range of 0.76–0.91 fm.[159]

On the other hand, according to Comay's model the charge radius of Σ^+ should be significantly larger than QCD predicts, falling in the

---

[158] K. Nakamura *et al.* (Particle Data Group), J. Phys. G 37, 075021 (2010)

[159] P. Wang, D. B. Leinweber, A. W. Thomas and R. D. Young, *Chiral Extrapolation of Octet Baryon Charge Radii,* Phys. Rev. D 79, 094001 (2009).

range of 0.91-1.12 fm.[160] The basics of the calculation of charge radius are quite simple. See Figure 36:

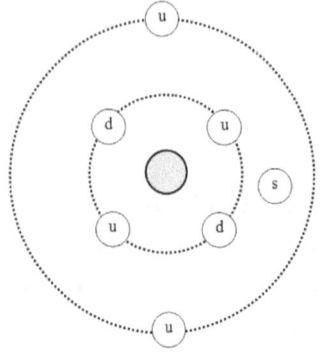
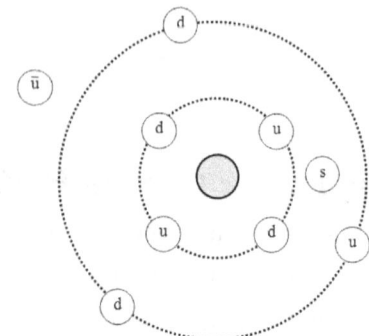

Figure 36. A typical configuration of Σ^+ baryon

Figure 37. A typical configuration of Σ^+ baryon with quark-antiquark pair

According to Comay's model, the *s* quark is closer to the baryonic center for two reasons: first, it is not repelled by the Pauli exclusion principle associated with the inner *u,d* quark shells since it has a different flavor and, second, it is heavier than *u,d* quarks and therefore is much closer to the baryonic center.

Assuming that like the proton, Σ^+ has additionally nearly half quark-antiquark pair (see Figure 37), the radius of Σ^+ should be the average between the configuration with a pair and the configuration without a pair. Below we can see a typical quark-antiquark pair in Σ^+. As we already know pairs of *d$\bar{d}$* are more likely (because of the *u* quarks) and the *$\bar{d}$* will be located in the periphery. This will slightly increase the charge radius of Σ^+ baryon.

Similar considerations may be applied to Σ^-.

---

[160] E. Comay, *Predictions of High Energy Experimental Results,* Prog. in Phys. **4,** 13 (2010)

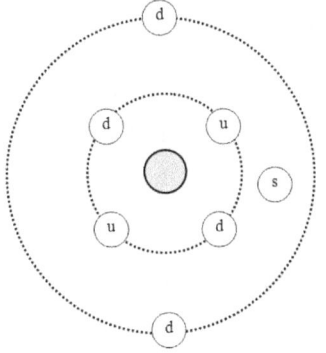

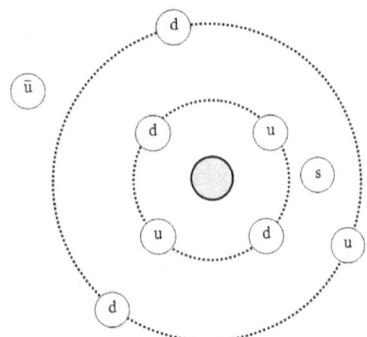

Figure 38. A typical configuration of
Σ^- baryon

Figure 39. A typical configuration of Σ^-
baryon with a quark-antiquark pair

The typical configurations of the proton are:

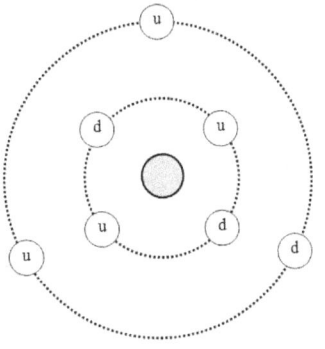

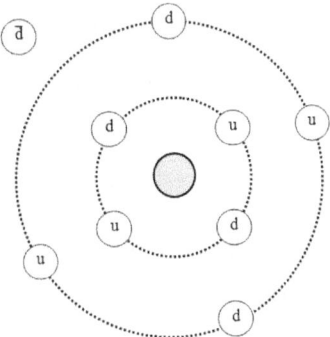

Figure 40. A typical configuration of
the proton

Figure 41. A typical configuration of a
proton with quark-antiquark pair

Calculating the charge radius involves summing the quarks' charge multiplied by the mean square distance from the baryon's center. This result is then divided by the total particle charge.

In the Σ^- baryon each of its *dds* valence quarks carry the same negative charge of $-1/3$. Hence, the smaller radius of the negatively

159

charged *s* quark is the main reason for the fact that the Σ^- baryon's charge radius is smaller than that of the proton.

In the case of the Σ^+ baryon, the result is reversed, because the total charge of Σ^+ is positive. Here the smaller radius of the *s* quark is the main reason for expecting a larger baryonic charged radius for the Σ^+.

Therefore, it is not hard to see that in Comay's model the following relations hold:

$$R(\Sigma^+) > R(\text{proton}) > R(\Sigma^-)$$

Where R denotes the particle's charge radius. In fact, according to Comay, in a first approximation, the proton charge radius should be rather close to the average between Σ^+ and Σ^- baryons. Comay's prediction cannot be more accurate mainly because of the inaccuracy of the measurements of Σ^- conducted so far.

The lattice QCD prediction is

$$R(\Sigma^+) \sim R(\text{proton}) > R(\Sigma^-)$$

Let us wait and see the experimental value of the Σ^+ charge radius and the comparison between the predictions of QCD vs. Comay's predictions.

UNIT 5: MAGNETIC MONOPOLES

Chapter 18: Magnetic Monopoles

Magnetic and electric forces are present in nature in a sort of "duality": electricity in movement creates a magnetic field, and a magnet in movement creates an electric field. They have other "dual" features as well. Up to now, electrically charged particles have been observed with either a positive or a negative electric charge. Magnets, on the other hand, only appear with 2 indivisible poles called "north" and "south" with respect to the direction of the Earth's magnetic field.

The Dirac monopole

In 1931, Dirac tried to predict the behavior of a "magnetic monopole", a single-pole particle never observed before. Particles have a definite electric charge – positive or negative (for example, the electron has a negative charge). Dirac was trying to predict the behavior of a particle carrying the dual charge – the magnetic charge.[161]

Without being explicitly aware of it, Dirac made the assumption that the ensemble of the properties of an electric charge-based electromagnetic field as science knows it, would be identical for fields generated by magnetic monopoles. When reading the article,

---

[161] The terms magnetic charge, magnetic monopole and monopole are synonyms.

some 50 years after it had been written, Comay noticed that Dirac was making this assumption even though it wasn't supported by any experimental evidence. Comay recalls that in his very first reading of this important article, he insolently wrote down an "X" next to the paragraph containing this hidden assumption.

In this article, Dirac preserved with the development of his theory, up to a point where he ran into a hard-to-crack mathematical glitch. In order to crack it, he invented a concept called "String", totally foreign to the established theoretical structure of the electromagnetic theory. Dirac's string defines a one-dimensional curve on which the electromagnetic equations become irregular[162].

Experimental physicists were trying to track down Dirac's monopoles for many decades with no success. Dirac himself continued to work on theoretical aspects of his monopole equation for a long time. In 1948, he made yet another attempt to derive his monopole equation from the Variational Principle[163], for which he had to make some additional heavy assumptions. The paper he published did not lead to the modification of experimental techniques, thus perpetuating the failure of the experimental endeavors to discover monopoles.

The strong interaction – is it magnetic force?

When trying to understand the source of the strong interaction which holds the quarks together, physicists wondered if it was a totally new

---

[162] The "String" idea spread out to a point where it now constitutes a research field in itself called "String Theory", relating to topics in physics very far from Dirac's monopoles.

[163] The Variational Principle works when a physical solution is a minimum or maximum of a certain function, e.g.: light moves in a straight line, which is the shortest line between two points.

and different force, or if it derived from forces they already knew, the electric and magnetic forces. It was easily shown that it could not be an electric force, but could it perhaps be a magnetic force? Could it be created from those magnetic monopoles described by Dirac, expected to be found within the proton and the neutron?

Physicists who tried to examine this idea, like for example Nobel Prize laureate Julian Schwinger, assumed that magnetic monopoles should follow the equations derived by Dirac in his 1931 paper.

Based on these equations, if the strong interaction is a magnetic force, the electrically charged electron should feel its influence when colliding with a proton or a neutron. Moreover, in this case the neutron's spin should produce a quite strong electric dipole field. Experimental findings refuted this idea and scientists therefore ended up believing that the strong interaction is not magnetic, but rather some other kind of force acting between particles, in addition to the electromagnetic force[164].

Comay's Monopoles

In 1982, during his stay in the University of Michigan, Comay came across a journal article describing a sensational experimental finding – a magnetic monopole. This result turned out to be wrong soon after, but it aroused Comay's curiosity and he started reading articles that discussed monopoles. His readings lead him to Dirac's monopole strings, which he found somewhat curious. He therefore decided to go back to the source – to Dirac's 1931 magnetic monopole paper. As we mentioned above, Dirac made a latent assumption in this article. Comay decided to check whether monopoles can be described without this assumption, and derive the

---

[164] J. Schwinger, *Sources and Magnetic Charge*, Phys. Rev. 173, 1536, 1968

equations merely from the "variational principle." This principle constitutes one of physics' solid foundations.

Within a few weeks, Comay developed the corresponding equations. If his equations are correct, then quarks are indeed magnetic monopoles, protons and neutrons are composed of such monopoles, and the strong interaction is nothing but a magnetic monopole force. In particular, Comay's equations explain why an electrically charged electron does not interact with the magnetic monopoles when colliding with a proton or a neutron.

This is a brief summary of the results derived from Comay's equations:

- An electric charge does not interact directly with a magnetic monopole.

- A photon interacts with both electric charge and magnetic monopole.

- A spinning electric charge creates an *axial* magnetic dipole (according to Maxwell equations).

- A spinning magnetic monopole creates an *axial* electric dipole (which is dual to the previous point).

- A magnetic monopole interacts with another magnetic monopole. It interacts with a *polar* magnetic dipole but not with an *axial* magnetic dipole.

- An electric charge interacts with another electric charge. It interacts with a *polar* electric dipole but not with an *axial* electric dipole (which is dual to the previous point).

- The elementary unit of the monopole is a free parameter.

166

- The theory is free from string irregularities.

These results show a complete duality between electric charges and magnetic monopoles. The difference between Comay's monopoles and Dirac's monopoles is that according to Comay, a magnetic monopole does not interact directly with an electric charge. Now, if we assume that quarks carry a magnetic monopole, this explains why an electron (which has both spin and an axial magnetic dipole) does not interact strongly with quarks.

The similarity between the electric force and the strong interaction has already been studied in the late 1960s by the physicist Phil Yock from New-Zealand. The papers he published on this subject remained without response from the scientific community. Comay went further and developed the mathematical infrastructure explaining the monopoles' properties, showing that the strong interaction can very well be a magnetic force carried by monopoles.

Overwhelmed by the scope of this discovery, he told one of his colleagues at the University of Michigan about it, who proposed to hold a seminar on the subject. Local monopole experts attended the seminar, manifesting skepticism. Several physics journals were also skeptical about this discovery and refused to publish it. It ended up being published in a physics journal almost a year later.[165] This theory is called Regular Charge-Monopole Theory (RCMT).

The publishing of a discovery does not necessarily lead to a shift of perception among professionals in the field. They are hardly to blame. There are currently some 300 new theoretical physics papers

---

[165] E.Comay, *Axiomatic deduction of equations of motion in classical electrodynamics,* Il Nuovo Cimento, **80B**, 159 (1984). www.tau.ac.il /~elicomay/nc84.pdf

published daily. New "discoveries" are constantly being published and taking the trouble to verify which of them is correct would require more than a lifetime. Therefore, due to this abundance of papers published every day, it is natural that only papers predicting an effect which ends up being experimentally verified would attract the scientific community's attention and be more carefully analyzed.

About the monopole equations

Why is it that this was not discovered before? Indeed, all it takes is grasping the deep significance of the variational principle. This understanding developed and was consolidated some 90 years ago. One can argue then that it should have been obvious, that every physicist should have thought of this model in the first place. But going this way meant parting from Dirac's 1931 theory, and it is likely that physicists' attachment to Dirac's approach to the monopole problem prevented them from pursuing the development of the idea.

Furthermore, Comay showed in 1995 that the magnetic monopoles equations could be derived even without an explicit utilization of the variational principle – by assuming the duality between the magnetic and electric fields.[166] The fact that different theoretical approaches lead to the same monopole equations is another reason to believe that these equations are correct.

---

[166] E. Comay, *Charges, Monopoles and Duality Relations*, Il Nuovo Cimento, Vol **110B**, N.11, 1995. tau.ac.il/~elicomay/nc95.pdf

Chapter 19: Comay's Monopoles at a Glance

Before we explain Comay's discovery, let's talk very briefly about a few terms well-known in the electromagnetic theory: the electric charge, the magnetic axial dipole, the electric polar dipole and the photon.

The **electric charge** is carried by massive particles such as the electron and the proton. Two electric charges with the same sign repel each other and two electric charges with opposite signs attract each other.

An **axial magnetic dipole** is created when electric charges move in a loop or when an electrically charged particle has a spin. Two dipoles apply forces on one another. The direction of the force depends on the exact setup of the dipoles and it generally does not coincide with the line connecting them. This kind of force is called a "tensor force".

A **polar electric dipole** is created when two opposite electric charges are positioned one next to the other and the sum of the charges vanishes. When observed from a distance, the electric charges of two objects nearly cancel each other out and they behave similarly to an axial dipole.

The basic unit of light, the **photon**, interacts with electric charges. The concept of the photon was developed by Einstein to explain the photoelectric effect. When a photon hits an electric charge, like the electron, it interacts with it, meaning that energy may be transferred between these particles. The intensity of the interaction depends on the photon's energy and on the mass and strength of the electric charge.

Comay's monopoles

Comay's monopole has a dual behavior with respect to the electric charge. Look at the table below:

Known electric terms	Comay's new terms
Electric charge	Magnetic charge, monopole
Axial magnetic dipole	Axial electric dipole
Polar electric dipole	Polar magnetic dipoles
Photon	

According to Comay's formulas, the set of well known objects in the left column does not interact directly with the set of objects listed on the right column. However, they both share the same radiation particle, the photon and interact with it.

As mentioned earlier, these properties were derived from very well-known and established physical ideas: special relativity, the Maxwell equations, the Lorenz force and the variational principle used for formulating the system's equations of motion.

Quarks as magnetic monopole carriers

When Comay started to develop these equations back in 1983, he wasn't thinking about the quarks because according to common scientific knowledge the idea of quarks being monopoles was not considered possible. His main goal was to find a theory capable of describing the monopoles without the irregularities that Dirac introduced into his theory, namely without Dirac's strings.

However, it turns out that the theory that Comay developed amazingly fits the behavior of the quarks. The strong force, according to Comay, is therefore carried by magnetic monopoles. We have already seen how his model fits many unexplained experimental results. We will see more in the next chapters.

The neutron's electric dipole

In the 1960s, Julian Schwinger looked into the possibility of quarks being magnetic monopoles. The fundamental monopoles' properties as tested by Schwinger were published in the 1931 article by Dirac. Measures showed that the upper limit of the neutron's electric dipole moment tends to zero. However, if quarks, whose spin=1/2, carry a magnetic monopole then their electric dipole moment should be very large.[167] Therefore, the neutron, composed of 3 quarks, is supposed to have a large electric dipole moment. This was probably the point on which Schwinger stumbled upon in his attempt to describe the quarks as monopoles.

---

[167] The proton's quarks establish a complicated system of spinning quarks in motion. The quark's electric charge is the origin of the proton's magnetic dipole moment. Now, these quarks also carry a magnetic charge which is stronger than their electric charge. Hence, the proton's axial electric dipole moment is expected to be stronger than its magnetic dipole moment. This argument also holds for the neutron.

171

On the face of it, the experimental fact showing that the neutron has no electric dipole moment should have totally discredited Comay's model as well.

But that's not exactly so. As Comay discovered, a magnetic monopole doesn't directly apply forces upon electric charges and this is the reason why the axial electric dipole associated with the monopoles does not exert a force on electric charges. Experiments measuring the neutron's electric dipole actually measured its interaction with electric charges, i.e., only the neutron's polar electric dipole was measured and was indeed found to be zero or tending to zero.

Comay's model therefore is perfectly coherent with these experimental measurements.

Chapter 20: Photons and Strong Interaction

Nearly 50 years ago, an experiment was conducted in which protons and neutrons were "bombarded" with highly energetic photons. The photons excited the proton and the neutron and consequently particles were emitted. The products of the collision of photons with protons were nearly identical to the particles coming out of the photon- neutron collisions. This was a surprising result.

Why did scientists expect that proton and neutron interactions should yield different results?

Based on the Maxwell Equations, developed in the 19th century that won the entire scientific community's global consensus, photons interact only with electric charges. At the time of the abovementioned experiment, scientists already knew that the neutron's electric structure was very different from that of the proton: the proton's charge is positive and the neutron has no electric charge, since the charges of its constituents cancel each other out. Today we know that the proton has two u-quarks of charge 2/3, and one d-quark of charge -1/3. The neutron, on the other hand, has two d-quarks and one u-quark.

If the nucleons are so different from each other with respect to their electric charges, how could energetic photon collisions with protons and neutrons lead to similar experimental results, given that photons are only supposed to interact with electric charges?

Another unexpected result was the intensity of the photon-proton interaction. The photon-nucleon interaction turned out to be much more intense than if it were merely an interaction between electric charges. One indication for the intensity of an interaction is the number of particles created by it, and in the case of the photon-nucleon interactions, this number was significantly higher than expected.

Vector Meson Dominance (VMD)

During the sixties, a young physicist named Jun John Sakurai published a theory in which he claimed that the photon is not merely a massless entity but is in fact a superposition of a "pure electromagnetic" photon and a "vector meson."[168,169] This theory, called Vector-Meson-Dominance (VMD), is the only current theory explaining the abovementioned effect. According to Sakurai, most of the hadrons coming out of photon-proton or photon-neutron collisions derive from the vector mesons present in the photons. The VMD assumption explains the similarity between the collision products of protons and neutrons.

---

[168] J.J. Sakurai, *Theory of Strong Interactions*, Ann. Phys. (NY) 11, 1–48 (1960) and also J.J. Sakurai, *Pion Resonances*, Nuovo Cimento 16, 388 (1960)

[169] A vector meson is a meson having a spin=1 and an odd parity, i.e., the ρ meson is a vector meson.

VMD's suggestion that the photon is a superposition of a massless electromagnetic photon and a massive hadron, contradicts special relativity, as Comay demonstrated.[170]

The scientific community is not entirely in peace with VMD either. As a matter of fact, VMD had been conceived before QCD and is still not a part of the standard model.[171] PACS, the Physics and Astronomy Classification Scheme, classifies the VMD concept as a model and not as a part of a theory. The nature of the interaction of energetic photons with nucleons has found no admitted explanation yet. Textbooks simply avoid mentioning the issue.

Nevertheless, the VMD idea is the only mainstream physics' attempted explanation for this phenomenon.

Comay's explanation for the photon-nucleon interaction

Comay's equations show that quarks, which carry a magnetic monopole as well as electric charge, do react to real photons. Furthermore, the magnitude of the quarks' magnetic charge is much larger than that of its electric charge[172]. Therefore, most of the interactions of photons colliding with protons or neutrons are the consequence of the quarks' magnetic charge, which is the same for

---

[170] E. Comay, *Remarks on Photon-Hadron Interactions*, Apeiron 10, No 2, 87 (2003).

[171] *"No direct translation between the Standard Model and VMD has yet been made."* H.B. O'Connell, B.C. Pearce, A.W. Thomas and A.G. Williams, *Rho-omega mixing, vector meson dominance and the pion form-factor*, Prog. Nucl. Part. Phys. **39** (1997) 201-252. arxiv.org/ PS_cache/hep-ph/pdf/9501/9501251v3.pdf

[172] The magnitude of the square of the quarks' magnetic charge is much larger than that of the electric charge, about 100-fold.

both *u* and *d* quarks, and are therefore the same for protons and neutrons.

The interaction of an energetic photon with hadrons provides a decisive test case in favor of Comay's model. In fact, the photon's interaction with the quarks is one of the most important building blocks of the theoretical foundation of Comay's model.

Chapter 21: The Three Jet Event

An interesting experiment of colliding high energy electron-positron beams has begun in 1978. The experiment was conducted in the PETRA collider at DESY. The collisions occasionally lead to the mutual annihilation of these particles, producing a large amount of energy, which prompts the creation of a new particle-antiparticle pair moving in opposite directions. This newly created pair could sometimes consist of muon and antimuon, but in some cases a pair of quark-antiquark is created. This initial pair creates particle jets in the detector, because of the action of the strong interaction on the quarks.[173]

Based on QCD, scientists predicted that occasionally three jets could be created, the third jet originating from a gluon. And indeed, these collisions did occasionally result in three particle jets.[174,175]

In the literature, the fact that three-jet collisions are experimentally observed is considered as a major proof showing the existence of

---

[173] This kind of process is very common in particle physics experiments. It demonstrates the equivalence of energy and mass. The kinetic energy of the colliding electron-positron is converted into particles whose self mass is much larger than that of the electron.

[174] R. Brandelik *et al.* (TASSO collaboration) (1979). *Evidence for Planar Events in e+e- Annihilation at High Energies*, Phys. Lett. B **86.** p.243–249

[175] cerncourier.com/cws/article/cern/39747

gluons[176]. Since a gluon does not exist as a free isolated particle, it cannot be directly observed and its existence can only be deduced from indirect experimental evidence.

The bremsstrahlung radiation

"The bremsstrahlung radiation" is a well-known process in electromagnetism in which photons are emitted during interactions between electric charges. The bremsstrahlung radiation's intensity is proportional to the sixth power of the charge and is therefore relatively weak in the case of electric charges.[177] For this reason, the bremsstrahlung effect is stronger for an electron which moves near the nucleus of a heavy atom, interacting coherently with the protons.

In general, QCD can be described as a hybrid between electromagnetism and a highly complicated mathematical structure, called the Yang-Mills SU(3) group. QCD's gluon, among others, is conceived as the strong interaction's equivalent to the photon in electromagnetism. QCD scientists have therefore transposed the bremsstrahlung idea onto the QCD framework and used it to predict the emission of gluons, which occurred in PETRA.

In the light of Comay's model the bremsstrahlung radiation does lead to the emission of an energetic photon[178], because the quarks are magnetic monopoles, and the square of their magnetic charge is approximately 100 times stronger than that of the electric charge.

---

[176] en.wikipedia.org/wiki/Gluon

[177] Just to get an idea, the square of the electric charge is approximated to 1/137 according to the unit system commonly used for calculations. Therefore, the charge's 6th power would yield a very small number.

[178] In Comay's model the photon works in perfect duality for electric charges and magnetic monopoles. Since Bremsstrahlung Radiation creates a photon with electromagnetic charges, similar phenomenon exists for magnetic monopoles.

Therefore, a magnetic monopole related bremsstrahlung may take place, in a total analogy to the electric charge process in which a photon is emitted. Since the basic unit of the strong charge is much larger than that of the electric charge, the bremsstrahlung photons emitted by the monopoles are expected to have a much higher probability.

This is the advantage of Comay's explanation. It's been known for over 50 years that energetic photons are involved in strong interactions – as we well remember from that inadequate explanation of the VMD theory. It is precisely this kind of energetic photon that generates the third particle jet in the PETRA experiment.

This result flows naturally from what is already known: the original bremsstrahlung referred to photons, and we're dealing with photons here too; the original phenomenon is based on electric charge equations, and here we're talking about quarks which are magnetic monopoles, fulfilling equations which are dual to electromagnetism. Furthermore, energetic photons have been long-known to be involved in strong interactions. Therefore explaining this phenomenon would not require any new assumptions.

Chapter 22: The π^0 Decay

The particle π^0 is a composite of the quarks u and $\bar{u}$ or d and $\bar{d}$. This particle mostly decays into two photons. It was known in the 1960s that π^0 decays into photons much faster than the expected decay time obtained from theoretical calculations according to field theory.

The fact that this particle decays around 9 times faster than the prediction is considered major proof for the validity of the QCD theory.[179]

The creation of photons in the process of π^0 decay is considered as related to the electric charge of the quarks, but the electric charge of the quarks is not strong enough to cause such a quick decay of the π^0. The calculation is based on field theory that accurately predicted the decay of positronium (the bound state of an electron and a positron).

Textbooks claim that the reason for the fast decay is the existence of 3 colors. The use of 3 colors in QCD's equations yields that π^0 should decay 9 times faster.

---

[179] Michael E. Peskin and Daniel V. Schroeder, *An Introduction to Quantum Field Theory* (Perseus, Reading, MA, 1995) p. 676.

Comay's explanation of the fast π^0 decay

It seems that in order to support their convictions, scientists sometimes tend to ignore facts that do not go smoothly with their model.

First, according to field theory, any meson, including the π^0, contains a high quantity of additional quark-antiquark pairs. This means that the probability to find an additional pair at a certain point in time is very high. When such an additional pair exists in the π^0, the particle might decay but the additional pair will remain. This effect causes a significant reduction of the speed of π^0 decay and produces an additional complication in the calculation that was neglected.

The π^0 decay is the reverse process of that in which photons hit hadrons and create quark-antiquark pairs. In the case of the π^0 decay, the quark-antiquark pair disappears and a pair of photons is created instead. The calculation used by scientists assumed that the reverse process is purely an electromagnetic process that depends on the electric charge of the quarks.

As we mentioned earlier, the fact that photons participate in the strong interaction was known in the 1960s, although it didn't receive the right explanation. Therefore, the reverse process is that the photons are created by the magnetic pole of the quarks.

Therefore, the equations used to calculate the π^0 decay should use the magnetic charge instead of the electric charge of the quarks. The magnetic charge is much stronger than the electric charge.

The conclusion is that QCD's calculation of the π^0 decay was wrong from the beginning.

Chapter 23: The Deuteron Tensor Force

According to the Regular Charge-Monopole Theory developed by Comay, there is a full duality between magnetic and electric forces. In addition to their electric charge, the quarks carry magnetic monopole charges with equivalent properties to those of electric charges.

Therefore, just as an electrically charged, spin ½ electron creates an **axial magnetic dipole**, the magnetically charged spin ½ quark creates an **axial electric dipole**.

Based on the equations Comay developed, the quarks carrying the monopoles do not apply an electrical force directly on the electric charges. It is rather the force exerted by one axial electric dipole on another. That is to say, protons and neutrons can be considered as axial electric dipoles exerting tensor force on one another.

Why the deuteron looks like a rugby ball

The deuteron is a nucleus of a heavy hydrogen atom containing a proton and a neutron. On the face of it, one may expect that the geometry of the deuteron should be spherical[180]. However, it turns out that the deuteron's shape looks more like an American football

---

[180] Had the Nuclear Force potential depended only on the distance between these proton and neutron, the deuteron's geometry should have been spherical.

or a rugby ball. Such oval shape may imply the existence of additional force acting between the proton and the neutron inside the deuteron. The field line of this additional force is not parallel to the line joining the two nucleons, just like the force lines between two magnetic dipoles. Nuclear scientists call this force "the tensor component of the nuclear force".

The first attempt to explain the deuteron oval shape is attributed to the proton and neutron magnetic dipole moment. Quarks are electrically charged: u quarks carry a charge of 2/3 and the d quark - 1/3. These quarks move within the nucleon – the proton or the neutron. Both their motion and their spin create magnetic moments.[181] As explained above, these magnetic moments generate a magnetic dipole force, which behaves like tensor force between the proton and the neutron.

However, calculations showed that this force is significantly weaker than the nuclear tensor force.[182] Furthermore, the calculated sign of the tensor force resulting from the quarks' electric charge is opposite to what is observed experimentally. Therefore, the electric charge of the quarks and the associated magnetic dipoles cannot explain the rugby ball shape of the deuteron.[183]

The origin of the tensor force between the nucleons is considered an open question.

According to Comay's model, quarks are magnetic monopoles, i.e., carrying magnetic charges. The model assumes a duality between the magnetic and the electric fields, and the quarks, with their magnetic

---

[181] The magnetic moment is a well known property of the proton and the neutron.

[182] A. deShalit and H. Feshbach, *Theoretical Nuclear Physics* (Wiley, New York, 1974) p. 13

[183] S. S. M. Wong, *Introductory Nuclear Physics* (Wiley, New York, 1998). p. 65

monopole charge and spin, create an axial electric dipole moment when moving within the nucleon's space.

The forces here are much stronger than the electromagnetic forces resulting from the quarks' electric charge: the square of the quarks' magnetic monopole charge is about 100 times stronger than that of the electric charge of the electron. Therefore, the force induced by the axial electric dipoles of the proton and neutron is not negligible.

Since all the quarks carry the same magnetic monopoles, the proton and neutron's axial electric dipoles are very similar. Furthermore, the sign (or direction) of the Nuclear Tensor Force is well-known from nuclear physics literature and is based on the rugby-ball shape of the deuteron. It turns out that the sign of the nuclear tensor force coincides with the sign of the force between two equal axial dipoles.[184]

The development of Comay's model, the Regular Charge-Monopole Theory, is based on considerations which are totally independent of the specific geometrical structure of the deuteron. The fact that this theory accounts for both the existence of the nuclear tensor force and its sign provides an additional experimental proof of its validity and of its adequacy in explaining the strong interactions.

---

[184] *A Regular Monopole Theory and Its Application to Strong Interactions*, Published in "Has the Last Word Been Said on Classical Electrodynamics?" Rinton Press, NJ, 2004. tau.ac.il/~elicomay/ LastWord.pdf

Chapter 24: Conservation Laws in Comay's Model

There are several important physical laws called "Conservation Laws". The most famous of them are the laws of energy and momentum conservation and the law of angular momentum conservation. As a matter of fact, additional conservation laws apply.

Conservation of electric and magnetic charges

The conservation of the electric charge is an inherent part of the Maxwellian electrodynamics. According to Comay's model, the equations of magnetic charge systems are analogous to Maxwell's electromagnetic equations and should obey a magnetic charge conservation law. According to Comay's theory, quarks carry one unit of a negative magnetic charge and antiquarks have one unit of positive magnetic charge.

Magnetic charge conservation introduces a new constraint which is constantly validated by experiments. The experimental discovery of one single process infringing this law would be enough to refute Comay's model.

The baryon number conservation law

Experiments show that the baryon number is conserved during interactions. Comay's model explains this conservation law by assuming the existence of an inner core inside every baryon.

185

The baryonic core contains quarks as well as an additional particle carrying a positive magnetic charge. This particle is not a quark and it can be annihilated only by its antiparticle which exists inside the core of the antibaryons. Thus, the process of baryon-antibaryon annihilation admits the baryon number conservation law. In other words, the number of baryons is equal to the number of cores. Basic laws of electrodynamics prove that this number is conserved.

Some theories, called "theories beyond the standard model", assume the existence of processes in nature in which the baryon number is not conserved. One of these processes, described in the model of Georgi and Glashow[185], is called "proton decay", suggesting that the proton can spontaneously disintegrate into a pi meson and a positron.

Another idea challenging the universality of the baryon conservation law hypothesizes the existence of a physical object having both a baryonic number ½ and a lepton number of ½.[186,187]

Most of the theories allowing the non-conservation of the baryon number argue that the individual baryon and lepton number conservation can be violated, as long as the difference between them is conserved.[188]

---

[185] Howard Georgi and Sheldon Glashow, *Unity of All Elementary-Particle Forces*, Physical Review Letters, **32** (1974) 438

[186] Klinkhamer, F. R.; and Manton, N. S. (1984). *A saddle-point solution in the Weinberg-Salam theory*, Phys. Rev. D **30**: 2212–2220

[187] The effect of this quite strange paper is very far from being negligible. Thus, as of November 2010, the paper containing this claim has been cited by 698 scientific articles.

[188] This is called B-L conservation law. Most of the theories assume that this number is conserved in any process. See en.wikipedia.org/ wiki/B-L

In spite of the great experimental efforts to find proton decay processes, no phenomena of baryon or lepton number conservation violation have been observed,[189,190] in line with the predictions of Comay's model.

The issue of the proton's decay is currently considered as one of the major open theoretical questions in physics.

The strong CP problem

Parity can be defined as symmetry between left and right. A physical process conserves parity if it occurs in an identical manner in a "mirror world", that is a world in which right and left are inverted (this symmetry is symbolized by the letter "P"). Maxwell's equations imply that the electromagnetic interactions do conserve parity. On the other hand, it has been known since the 1950s that weak interactions do not conserve parity.

The symmetry called "charge conjugation" (symbolized by the letter "C") means that two closed identical systems differing from each other only by the sign of their charge,[191] should have identical behavior.

The original formulation of QCD was supposed to conserve C, P and their combination CP. But QCD belongs to a global theory, the standard model, which imposes non CP conserving hypothetical processes.[192] The standard model introduced a new parameter called

[189] B. V. Sreekantan, *Searches for Proton Decay and Superheavy Magnetic Monopoles*, Journal of Astrophysics and Astronomy **5**, (1984) p.251–271

[190] H. Nishino *et al. Search for Proton Decay via* $p \rightarrow e+ \pi^0$ *and* $p \rightarrow \mu+ \pi^0$ *in a Large Water Cherenkov Detector*, Physical Review Letters **102**, 14 (2009)

[191] In the other system every particle is replaced by its antiparticle.

[192] I.I. Bigi and A. I. Sanda, *CP violation* (Cambridge, University Press, 2000). p. 269

"theta" into the QCD equations, and on the basis of various theoretical considerations the value of theta has been evaluated as approximately 1. According to the QCD equations, if this parameter is different than 0, then parity conservation is violated in processes ruled by the strong interactions. Therefore, according to the standard model, CP conservation under the strong interaction should be violated.

All the experiments conducted thus far have shown, at a precision level of at least 10 decimals, that the strong interaction, like the electromagnetic force, does conserve parity and charge conjugation. These experimental results raise a problem called in physics the "strong CP problem".

The strong CP problem is considered as one of the major unsolved problems in particle physics.

According to Comay's model, the strong interaction is a magnetic monopole force, obeying equations dual to the electric charge force equations. Processes involving strong interactions are therefore expected to conserve both parity and charge conjugation. Comay's model is not a part of the standard model and is therefore free from its external constraints. In fact, any experimental finding indicating that the strong interaction violates parity, as predicted by the standard model, would collapse Comay's model.

It turns out that in addition to the long list of QCD contradictions, it has had to adapt itself to the standard model, and these adaptations and constraints sometimes lead to peculiar predictions (like strong CP violation) which have no grounds in reality.

UNIT 6: SUMMARY

Chapter 25: And Yet, Why Do Scientists Believe in QCD?

In this chapter we present the most convincing arguments supporting QCD that we collected consulting QCD experts and physics literature, as well as counter-arguments as regards Comay's theory.

Is QCD as solid a theory as the special relativity and quantum field theory?

An equation *predicting* a specific phenomenon with a high level of accuracy is likely to convince us of the validity of the theory behind it. The Dirac equation, for example, predicts certain phenomena with an accuracy of 4-6 decimals. The quantum field theory of the Dirac equation has increased this accuracy in many cases up to 9-10 orders of magnitude. This level of predictability is considered decisive evidence for the validity of these theories.

However, the QCD Wikipedia entry (of March 2011) reveals the following fact:

"Quantitative tests of non-perturbative QCD are fewer, because the predictions are harder to make. The best is probably the running of the QCD coupling as probed through lattice computations of heavy-quarkonium spectra. There is a recent claim about the mass of the heavy meson Bc. Other non-perturbative tests are currently at the level of 5% at best."

193

In many cases it appears that QCD is not even close to two digit accuracy. In the past, theories with much higher predictability, such as the Bohr-Sommerfeld atomic model, were abandoned for better theories.

So why are experts so certain that QCD is correct?

Phenomenological evidence

There are several phenomena that according to QCD experts can be explained only by QCD.

The properties of Ω⁻. Experts claim that quantum mechanics cannot explain this particle without using QCD colors. As we mentioned previously[193], this argument is incorrect. Not only that, but there are known particles in nature which have similar properties without colors. If this argument was correct, then why do such particles exist without colors?[194]

The three jet event[195]. According to QCD experts, in some electron-positron collisions three jets should be created, in which the third jet is initiated by a gluon. They measured its spin and it was 1. They also found that it interacted strongly with other hadrons.

The experts did not consider the possibility of the third jet being a photon[196] arguing that a photon interacts only with electromagnetic charges and therefore cannot explain the strength of the interaction of the third jet. Apparently the significance of the hadronic properties of the photon has eluded them[197], as if the interaction of

---

[193] See chapter "Wigner, Racah and the Three Quarks."

[194] See chapter "Non-Conservation of Knowledge" section "The NIH syndrome."

[195] See chapter "Three Jet Event."

[196] The photon also has spin-1.

[197] See chapters "Photons and Strong Interaction"

photons with hadrons is "filed" under another field of research. Those who have devoted their entire scientific career to QCD might not be aware of the importance of strong interaction aspects of the photon.

π^0 **decay**[198]. This is also considered proof, and just as in the case of the three jet event, the explanation based on the well established fact of the hadronic properties of the photon was not even considered.

Landau Pole. In 1955 Lev Landau and his co-workers discovered that in some cases the quantum electrodynamics (QED) equations become invalid.

Landau himself didn't consider this to be a problem of QED. The problem doesn't appear in the strong interaction according to QCD.

It is hard to understand why experts consider this fact as proof of QCD validity. Since the "problem" is not settled in electrodynamics, it is therefore reasonable to assume that this is indeed not a problem at all, as Landau claimed.

Quark confinement. The fact that no free quark was ever found is considered as proof for the validity of QCD. Both QCD and Comay's model provide explanations to the confinement phenomenon.[199]

Hadrons with 2 or 4 quarks do not exist. This is also considered as a proof of QCD validity. The inexistence of hadrons with 2 and 4 quarks is explained in Comay's model as well. Furthermore, QCD conceives the existence of other particles such as pentaquarks and

---

[198] See chapter "π^0 decay"
[199] See chapter "Confinement and Asymptotic Freedom"

other quark combinations that have not been found experimentally, exactly as Comay predicted.[200]

Lattice QCD

QCD equations are considered very complex. Ken Wilson, the 1982 Nobel Prize laureate, introduced the computational tool of Lattice QCD to facilitate the solving of these equations. QCD numerical calculations require extremely heavy computational resources, even when using lattice QCD, and can only be performed by the most powerful supercomputers available.

By means of lattice QCD, scientists were able to verify many experimental results at a precision level of 5% or less. However, even QCD experts themselves admit that lattice QCD did not issue any significant successful predictions, qualifying it as "Nothing to write home about."

Predicting hadronic mass and radius

One of the arguments considered as proof for the validity of QCD is its ability to predict masses of new baryons and mesons. In order to predict the mass of a new particle, scientists take known baryons and mesons with similarities to the predicted particle and use the experimental quantities to predict the new particle properties.

However, such method was quite successful even in the 1960s, developed by Gell-Mann[201] and Okubo[202] and called the Gell-Mann–Okubo mass formula[203].

---

[200] See chapter "QCD's predictions".

[201] M. Gell-Mann, *The Eightfold Way: A Theory of Strong Interaction Symmetry*, California Institute of Technology Synchrotron Laboratory Report CTSL-20 (1961)

The Gell-Mann–Okubo mass formula matches measured hadron masses, nearly as accurate as QCD predictions. It was invented many years before QCD, and it doesn't rely on it. It was even improved later[204].

However, the prediction of hadrons' charge radius is a different story. Unlike the mass predictions, a good prediction of a particle's charge radius often requires deeper understanding of its structure and of the interaction between its constituents. To date, we know of one prediction of the charge radius of Σ^+ based on lattice QCD.[205] This yet unverified prediction is new (2009) and we believe that experiments will show that it is completely off.[206]

QCD and the strong nuclear force

As we have tried to show in this book, QCD does not provide a satisfactory explanation of the strong nuclear force. The 2004 Nobel Prize laureate, Frank Wilczek, who received the prize for a central pillar of QCD, the asymptotic freedom, wrote a very interesting article in 2007 about the issue of QCD and strong nuclear force.[207]

Wilczek begins his article by admitting that he had always thought that QCD was incoherent with the experimental facts we all know

[202] S. Okubo, *Note on Unitary Symmetry in Strong Interactions*, Prog. Theor. Phys. 27 (1962)

[203] en.wikipedia.org/wiki/Gell-Mann–Okubo_mass_formula

[204] E.g.: *Gell-Mann Okubo Mass Formula Revisited*, L. Burakovsky and T. Goldman, 1997. arxiv.org/abs/hep-ph/9708498

[205] P. Wang, D. B. Leinweber, A. W. Thomas and R. D. Young, *Chiral Extrapolation of Octet Baryon Charge Radii,* Phys. Rev. D **79**, 094001 (2009)

[206] Eliahu Comay, *Predictions of High Energy Experimental Results, Progress in Physics*, Vol. **4**, 2010. p.13-16. See also the chapter "The charge radius of Σ+ baryon".

[207] Frank Wilczek, *Hard-core revelations*, Nature, Vol **445**, 156 (2007).

about strong nuclear force: "Ironically, from the perspective of QCD, the foundations of nuclear physics appear distinctly unsound."[208] He continues by explaining what the contradiction is: "Yet QCD tells us that protons and neutrons are themselves built from quarks and gluons that move at very nearly the speed of light. These more basic particles carry colour charges, leading to the additional requirement that they be confined within 'bags' whose contents are overall colour-neutral." And he further asks: "But why don't the separate proton and neutron bags in a complex nucleus merge into one common bag? On the face of it, the one-bag arrangement has a lot going for it. It would allow quarks and gluons free access to a larger region of space, and so save on the energetic cost of localizing their quantum-mechanical wave functions. But in such a merger, protons and neutrons would lose their individual identities, and our traditional, quite successful model of atomic nuclei would crumble. What prevents that calamity?"

Frank Wilczek raises one of the main problems of QCD.

In order to explain the properties of Δ^{++} and Ω^-, QCD theorists invented the colors, which allow quarks with exactly the same properties but different colors to live in the same space. But by doing so, a much more serious problem appeared: why don't proton and neutron merge and produce a particle with three u and three d quarks? This QCD invention is incoherent with the measurement that the atomic nuclear density is constant and the deuteron is a very loosely bound proton-neutron state.

---

[208] This is also claimed by Wong, the nuclear physicist. S. S. M. Wong, *Introductory Nuclear Physics* (Wiley, New York, 1998). p.102

The continuation of Wilczek's article is even more interesting. It discusses another article, written by Ishii, Aoki and Hatsuda[209] that attempts to explain this problem using lattice QCD. However, the required results *cannot* be obtained using the known meson properties, as the authors themselves admit. They can be explained by substituting experimentally well established meson masses with different and completely unphysical masses that are used as parameters to the lattice QCD engine. This is completely inconsistent with the experimental data of meson mass. For example, they used a pion's mass of 0.53 GeV whereas the true value is approximately 0.14 GeV.

It can therefore be concluded that the very serious problems of QCD regarding the strong nuclear force are still unsettled.

---

[209] N. Ishii, S. Aoki and T. Hatsuda, *The nuclear force from lattice QCD*, Phys. Rev. Lett. 99 (2007)

Chapter 26: Theories vs. Experimental Findings

The strong interaction as it is conceived by Comay's model and by QCD is fundamentally different. Comay's model is based on the Regular Charge Monopole Theory and describes every nucleon as having a massive core attracting quarks to it, in an analogy to the attraction exerted on electrons by the atomic nucleus. The QCD nucleon has no core, and an attractive force operates between the quarks. Let's try to find out which of these theories is more plausible.

QCD provides no explanation to some aspects of the similarity between the nuclear force and the van der Waals force, as shown by the potential curve of these forces. QCD explains neither the constant density of nucleons in nuclei nor the first EMC effect. All these phenomena flow naturally and straightforwardly from Comay's model.[210]

In order to explain the residual nature of the strong nuclear force, QCD further suggests that the force acting between the quarks inside the nucleons within the atomic nucleus arbitrarily ceases at a certain distance, without explaining why. Comay's model, on the other hand, naturally explains this phenomenon by the screening effect,

---

[210] See chapter "What Theories Say."

which is a well known feature of electromagnetically neutral systems.

QCD offers no explanation to the strong interaction's fundamental characteristics, such as the intensity of the interaction between energetic photons and nucleons, and the similarity between the behavior of protons and neutrons when they are hit by an energetic photon. Comay's model provides an immediate explanation to these phenomena, describing the quarks as magnetic monopoles obeying the RCMT equation of motion.[211]

Furthermore, it is not clear how QCD could explain the strength and the sign of the deuteron's tensor force. Comay's model provides an explanation to that phenomenon as well, considering the quarks as magnetic monopoles.[212]

QCD offers no explanation for the antiquark's peripheral location inside the nucleon, or to the fact that negative charges tend to be located at the neutron's external regions. Both these features flow directly from Comay's model describing the attraction-repulsion forces between the quarks, the antiquarks and the nucleonic core[213].

Furthermore, QCD's explanation to the confinement effect seems to be problematic, since in a situation where "everyone attracts everyone", it is unclear why mesons are not confined inside the proton.[214]

QCD's asymptotic freedom property cannot explain the slow decrease in the proton-proton cross section curve. This curve shows

---

[211] See chapter "Photons and Strong Interaction"

[212] See chapter "The Deuteron Tensor Force"

[213] See chapter "Further Evidences for Repulsive Forces in the Proton"

[214] See chapter "Confinement and Asymptotic Freedom"

that the forces between quarks of incident proton and quarks of target proton increase as the distance between them decreases.[215] Furthermore, QCD's asymptotic freedom cannot explain the proton form factor and the pion form factor that provide clear evidence that the forces between quarks inside hadrons decrease while the distance increases.[216]

QCD does not provide any explanation for the findings suggesting that nucleons have a non negligible core.[217] The increase in the total and the elastic cross-section curves of collisions between energetic protons, measured during the last decade, implies that there is a non trivial core inside the proton. In particular, the increase of the *elastic* cross section indicates that the proton contains a solid inner object that can take the heavy blow of the energetic collision and still keep the two protons intact. Comay's model provides an immediate explanation to these phenomena. The existence of a massive core within the proton is further supported by the ratio between the radii of corresponding baryons and mesons, and the mass ratio of corresponding baryons and mesons. [218]

QCD has predicted the discovery of several particles and entities such as strange quark matter, pentaquarks, dibaryons and glueballs, which have never been found in spite of great endeavors over the

---

[215] See section "The P-P cross section graph vs. asymptotic freedom" in chapter "Confinement and Asymptotic Freedom"
[216] See section "Proton form factor" in chapter "Confinement and Asymptotic Freedom"
[217] See chapter "Something Else Inside the Proton"
[218] See chapter "Baryon Radius too Large".

past few decades. Comay's model implies that these objects simply do not exist[219].

QCD, as part of the standard model, predicts strong CP violation which has not been observed. Comay's model explains why CP violation cannot occur in strong interactions[220].

QCD consolidates the quantum states of the Ω^- and the Δ^{++} baryons with the Pauli exclusion principle by inventing the colors, thus adding a new degree of freedom to the quarks' dynamics[221]. In so doing it ignores the existence of multiple configurations inside the proton that explain these particles' state by ordinary quantum mechanical laws.[222,223] Furthermore, QCD fails to account for the proton spin crisis which can be resolved immediately by adopting a mathematically consistent model with multiple configurations.

Additional phenomena explained by Comay's model

Whereas other phenomena predicted by Comay's model do not contradict QCD, they naturally flow from the model without any additional assumptions, further supporting its validity.

And indeed, the experimental measure of mesons and baryons' radii and masses are compatible with Comay's model: pion's radius, for example, is smaller than that of the proton, K^+ is smaller than π^+, and Σ^- is smaller than the proton. And as we noticed earlier, the masses of mesons and baryons are consistent with Comay's model.

---

[219] See chapter "QCD's Predictions"

[220] See chapter "Conservation Laws in Comay's Model"

[221] See chapter "What Theories Say"

[222] See chapter "Wigner, Racah and the Three Quarks"

[223] See E. Comay, *On the Quantum Mechanical State of the Δ^{++} baryon*, arxiv.org/abs/1011.1610

Chapter 27: Theories Foundation

The rather long list of QCD's failures makes a solid basis for its rejection. Besides this aspect, one may use a milder kind of argument that relies on QCD's logical structure.

Back in the 14th century, the Franciscan monk William of Occam formulated an intuitive rule of thumb, "Occam's razor", for the comparison between theories aiming at explaining phenomena. According to this rule, one should always prioritize the simpler theory between the two.

Therefore, a model requiring a smaller number of assumptions, and providing explanation for a larger number of phenomena, is naturally more convincing. On the other hand, a theory based on a large number of adjustable free parameters, ready to be adapted to the discovery of new experimental facts, should naturally seem weaker.

Assumptions in Comay's model

Comay's model is based on several assumptions. It assumes that the Variational Principle is valid for magnetic monopoles. The Variational Principle has been known for centuries, and has been used in mechanics, electromagnetic theory and quantum mechanics long before Comay conceived his model.

According to Comay, the properties of magnetic monopoles derive from this principle in analogy with the Maxwell Equations and the Lorenz Force. Maxwell equations and the Lorentz force are admitted by the entire scientific community as cornerstones of electrodynamics.

Comay further assumes that quarks carry magnetic monopoles, obeying the Regular Charge Monopole Theory. In this theory, the elementary monopole unit is a free parameter. Comay assumes that this unit is significantly larger than the elementary electric charge unit.

Comay's model of strong interaction assumes the existence of a core inside the baryons, containing three units of magnetic charge, and that each of the quarks has a negative magnetic charge of a magnitude of one unit. In analogy to the structure of electrons inside the atom, Comay suggests that this core has inner closed quark shells.

These are the only assumptions Comay makes. They have a strong resemblance to the structure of atomic electrons.

QCD assumptions

Like Comay, QCD assumes the validity of the variational principle. However, in the case of QCD, the central expression called the Lagrangian density differs from that of Comay and uses the Yang-Mills SU(3) theory. QCD further assumes that the strong interaction is characterized by an SU(3) triple charge, designated as three colors, an unpaired phenomenon in physics.

QCD claims that the force between the baryonic quarks is a force of attraction. QCD analysis shows that this force increases as quarks move away from one another. This is contrary to any other known natural elementary force.

Furthermore, QCD states that this force ceases its action at a certain distance with no consensus theoretical explanation for this interruption[224].

An additional assumption, according to QCD, is that isolated particles must always contain an equal amount of the 3 colors (they are therefore called "white particles"). Particles which would contain an unequal amount of color cannot exist separately and cannot be measured by the instruments. This prohibition likewise has no equivalent in physics.

QCD further assumes that each hadron contains gluons which carry color and anti-color of a different type. This is unmatched to any other known force in nature.

Which theory is more natural

The most significant difference between Comay's model and QCD is in Comay's conception of the strong interaction as analogous to the electromagnetic interaction, whereas according to QCD, the strong interaction is a force of a new kind related to the Yang-Mills group SU(3). In order to define this new force, QCD founders made up new mathematical structures and equations.

Stating that the strong interaction has no parallel in nature granted QCD theoreticians with great liberty to invent new laws in order to retroactively explain experimental findings.

---

[224] Some physicists assume that Yukawa's theory explains this, and some believe that color forces cancel each other.

QCD developers took great advantage of this freedom. For example, when it was discovered that only one half of the proton's momentum is carried by known quarks, QCD conceivers attributed the missing momentum to the massless gluons, which are undetectable in direct measurements[225].

In spite of these degrees of freedom, QCD is still unable to provide an explanation for many findings quoted here, some of which seem to blatantly contradict the QCD theory.

Comay's model, on the other hand, is based on a small number of hypotheses, with very few degrees of freedom. As we've seen above, the model's basic assumption is that the strong interaction is analogous to the well-known electromagnetic force. According to Comay, the major difference between the electromagnetic force, to which electric charge is associated, and the strong interaction, to which magnetic monopoles are associated, is the order of magnitude of the basic charge unit. This implies that electromagnetic effects should have parallels in systems ruled by the strong interaction.

Therefore, it would only take the discovery of pentaquarks, strange quark matter or glueballs, or the showing by measurements that the K meson radius is larger than that of the pi meson, or even just the observation of a steeper decrease of the proton-proton cross section at high energies, or simply finding that the Comay's calculated sign of the nuclear tensor force is opposite to the tensor force experimental sign to contradict Comay's model. If only antiquarks were not pushed toward the proton's periphery, or if quarks did carry the entire proton momentum, or if strong interactions caused CP violation, Comay's model would be proven to be wrong.

---

[225] This finding also surprised QCD developers, see D. H. Saxon, *The Lepton-Hadron Interaction,* Proc. R. Soc. Lond. A 404, 233 (1986). p. 239

The validation of only one of these phenomena or many of the other supporting phenomena mentioned in this book, often breaching the wall-to-wall consensus of the physicists' community, would bring about the collapse of Comay's model.

So, was Comay's model particularly lucky, or does it present a valid physical theory after all?

Chapter 28: Concluding Remarks

One must start with errors and convert it into truth.

That is, one must reveal the source of the error, otherwise hearing the truth won't do any good. The truth cannot force its way in when something else occupying its place.

*To convince someone of the truth, it is not enough to state it, but rather one must find the **path** from error to truth.*

(Ludwig Wittgenstein, remarks on Frazer's Golden Bough)

One of the central questions which probably crossed the reader's mind is why does an entire scientist community, many of which are undoubtedly gifted and even brilliant, stick to a theory which is shown here to be filled with as many holes as a sieve. The unshakable confidence of the scientific community in QCD greatly intensifies the reader's faith in it. Therefore, the historical explanation plays a significant role in understanding this situation, as we have tried to illustrate here.

In this matter, it is important to understand that the knowledge we gather about the particles composing the proton increases as particle accelerators reach higher energies. In the 1930s protons and neutrons were considered to be elementary particles, and Yukawa's theory was meant to explain only one rapidly decreasing attraction force.

QCD was born in the late 1960s and the inventors of QCD had basically 3 fundamental erroneous assumptions:

- They assumed that nucleons are composed of three quarks without any additional massive baryonic core. This seemed to be a natural assumption that they made immediately after the quarks were confirmed. They didn't know back then that a significant part of the mass is not carried by these quarks.

- They were surprised by the properties of Ω^- and Δ^{++} because they were not aware of Wigner and Racah's theory and its possible application to the quarks.

- They were engaged with Dirac and Dirac-like monopole theories, which could not be applied to quarks.

A totally different question, though, is how come this theory has remained unshakeable during the last 40 years in spite of its inconsistency with so many different experimental findings. This question is interesting, but analyzing it would certainly be beyond the scope of this book.

Predictions

In view of this comparison between QCD and Comay's model, one could think that the scientific community would transfer all the QCD scientific literature to science's history departments. But past experience shows that scientific revolutions are usually a long, hard and painful process.

Comay founded his theory on solid mathematical developments which would convince theoretical physicists in the field. Nevertheless, Comay suggests several experiments, the results of which would be decisive for determining the predictive capacity of each model and therefore the validity of the theory behind it.[226]

The most decisive experiment would be to scatter electrons on π^+ or π^- mesons in order to determine the mass distribution of the quarks inside the pion.

According to Comay, the mass will be found exclusively in the quark-antiquark (u and $\bar{d}$ in the π^+ case) and in the additional quark-antiquark pairs (that should exist in pions as they exist in nucleons). According to QCD, there are gluons inside the hadrons, carrying a significant amount of the mass (half of the mass in the case of the proton, and it should be similar in the case of π^+ and π^-.) Therefore, the results of such experiment will necessarily contradict at least one of these competing theories[227].

Another interesting experiment would be producing the pion-pion cross section curve at very high energies. Comay predicts that unlike the proton-proton collisions, in the case of pions the graph would decrease entirely. This is simply because pions have no inner quark shells that may enter high energy interaction as in the case of the protons.

Epilogue

The theoretical establishment of Comay's model began in the formulation of the magnetic monopoles properties in 1983. On this basis, he found the correspondence between experimental results and these quark properties.

---

[226] E. Comay, *Predictions of High Energy Experimental Results*, Prog. In Phys., Vol **4**, 13 (2010)
[227] See chapter "A decisive experiment"

During the 1980s Comay gave a seminar about his monopole theory before his colleagues at the University of Michigan in Ann Arbor. Another seminar, which also described applications to strong interactions, was given at the Tel Aviv University. In his lectures he described the early results of his work on monopoles.

The physics and the mathematics of the Dirac magnetic monopole theory are far from trivial. In particular, the mathematical structure of his monopole theory is very complicated and totally different from that of the well known Maxwellian electrodynamics of electric charges. Monopole theory drew the attention of a few other physicists who introduced modifications into Dirac's monopole theory while conserving the theory's main ideas and the incompatibility with Maxwellian electrodynamics of charges.

On the other hand, Comay has decided not to pursue Dirac's work on monopoles and to use instead the natural postulate where electrodynamics of monopoles without charges should be dual to the ordinary Maxwellian electrodynamics of electric charges that contain no monopole.

Dirac monopoles have been searched in vain for many decades. The consequence of this unsuccessful search was an unappealing reputation for the idea of magnetic monopoles which have led most physicists to staying away from them.

Comay did not realize that contradicting an apparently established theory goes way beyond producing a scientific work. While being positive of the validity of his model, he didn't understand how important a good marketing campaign was for making other scientists listen to him. They didn't, perhaps because he brought them the hard-core theory before attracting their attention to the scientific motivation behind its development. There were other reasons, however. During the 1980s he didn't have the whole picture as demonstrated in this book.

Personally, I believe that in a few decades from now, historians will likely look back at QCD development as one of the most hallucinating events in the history of science. I sincerely hope we will not have to wait so long.

APPENDICES

Terminology

In order to enhance the clarity of the text, we bring here in alphabetical order, definitions of some of the physical notions and concepts mentioned throughout the book.

Antiparticle: every elementary particle fulfilling the Dirac equation, has an anti-particle. The charge of the antiparticle is opposite to that of the particle.

Antiquark: the quark's antiparticle. There are 6 antiquarks, one for every quark.

Baryon: A particle family which among others includes the proton and the neutron. Except for the proton, when Baryons are free, they disintegrate, such that the final state includes a proton and particles whose Baryonic number adds up to zero. Every Baryon is characterized by a 3-quark combination having a specific quantum state.

Boson: a denomination for an integral (non-negative) spin particle. See also Fermion below.

Cross section: a useful notion for describing the outcome of an experiment where two particles collide (called a scattering experiment). The cross section for a specific outcome is simply proportional to the number of events that belong to this kind of results.

217

Dirac equation: quantum mechanics' fundamental equation for massive, spin 1/2 elementary particles. The equation successfully predicted major phenomena and finds a natural place in field theory. At its non-relativistic limit, and when ignoring spin effects, Dirac equation becomes the Schrödinger Equation.

Dirac particle: an elementary spin 1/2 particle. The leptons, quarks and their antiparticles are Dirac particles.

Electromagnetic force: the force applied by electric and magnetic fields on electric charges. Electric and magnetic fields are described by the Maxwell Equations.

Electron: an elementary particle present in the atomic shell. The electron has an electric charge of -e.

Elementary particle: a point-like particle. Elementary particle is not composed of any other particles.

Fermion: particles are classified into two categories based on their spin: particles with integral spin are called Bosons, and particles with spin equals to a non-negative integer plus one-half are called Fermions. Leptons, quarks, and nucleons are examples of fermions.

Gluons: physical QCD objects playing a role analogous to that of the electromagnetic field in the conventional atomic theory. However, the electromagnetic field also contains a radiation field (photons) emitted and absorbed by electric charge systems. On the other hand, there is no gluon radiation and that's why gluons have never been directly observed in any experimental device. This means that their existence is QCD dependent.

ħ: see- Planck's constant.

Hadrons: a general term designating particles composed of quarks (Baryons and Mesons).

Heisenberg uncertainty principle: this principle defines a lower limit to the imprecision related to physical quantities, formulated by means of ħ. This principle applies to the imprecision product of time and energy. For example, a particle whose typical lifetime is very short, allows, in principle, a good knowledge of the time interval during which it took part in the measurement. Therefore, its intrinsic energy measure (i.e. its mass) will be associated with a large inaccuracy. These relations also apply to position and momentum.

Isospin: A mathematical symmetry related to a system of protons and neutrons in a nucleus. It is based on the similar behavior of protons and neutrons in strong interactions. Its underlying basis is the similarity between the u,d quarks with respect to strong interactions. This symmetry is also found in systems of hadrons. It has the same SU(2) mathematical structure as that of spin.

Klein-Gordon Equation: a relativistic quantum mechanics equation supposedly describing a spin-0, massive particle.

Leptons: a family of elementary particles that contains the three types of electron, their corresponding neutrinos, and their antiparticles.

Liquid drop model: a model that makes an analogy between the nucleons inside a not very light atomic nucleus and molecules inside a liquid drop.

Maxwell Equations: Electromagnetic Theory's fundamental equations. These equations predicted the existence of electromagnetic waves and are perfectly coherent with Special Relativity.

Meson: a particle made of a quark and an antiquark.

Muon: a lepton similar to an electron that carries negative charge. The Muon mass is about 200 times larger than that of the electron.

Neutrino: a particle carrying no electric charge, of spin 1/2. There are 3 types of neutrinos, corresponding to the 3 members of the electron family: the electron, the muon and the tauon (called neutrino, muon –neutrino and tau-neutrino). Neutrinos were believed to be massless, but experiments conducted during the last decade suggest that they have a very small mass, although the possibility of their being massless is not totally discarded yet. The neutrino does not participate in Strong and Electromagnetic Interactions.

Neutron: a particle present inside the atomic nucleus. It is very similar to the proton, but its total electric charge is zero.

Nucleon: proton or neutron.

Parity: An important and useful property of some quantum mechanical systems like that of the electronic state of an atom, a nucleus, a meson and a baryon. Parity can be either positive (even) or negative (odd).

Pauli exclusion principle: states that two identical fermions are not allowed to be at the same quantum state. Mathematically speaking, the Pauli principle requires that the wave function of any two identical fermions is an anti-symmetric function of 2 orthogonal single-particle wave functions. The incompressibility of fluids and solids is the consequence of Pauli's principle as it is manifested in dense electron systems like that of liquids (and solids): a slight compression increases the electrons' probability to jump up to a higher energy level and the liquid's self energy increases. Since every system "tends" to minimize its energy, the liquid responds with a strong resistance to compression.

PDG: Particle Data Group: an international organization centralizing and processing particle related experimental results. This organization is considered as the oracle for establishing the existence of particle and determining their physical properties.

Photon: an elementary particle, the quantum of the electromagnetic radiation. It is the basic unit of light and of all other forms of electromagnetic radiation.

Pi meson (Pion): a meson composed of a quark and an antiquark from the u,d flavor of a specific quantum state (there are 3 types of pions). Pions are the lightest mesons.

Planck's constant: a very fundamental constant in Quantum Theory. It serves as a natural measure unit for the angular momentum and the spin of particles and quantum states. The original constant is marked with h. A more useful constant, called ℏ ("h-bar"), is $h/2\pi$.

Positron: the electron's anti-particle. It is identical to the electron in all aspects except for having a positive electric charge.

Positronium: a bound quantum state of an electron and a positron.

Proton: a particle inside the atomic nucleus. The proton has a positive +1 electric charge.

Quantum Field Theory (QFT): provides a theoretical framework for constructing quantum mechanical models of fields..

Quark: an elementary particle. There are 6 kinds of quarks: u, d, s, c, b, t nicknamed Up, Down, Strange, Charm, Bottom and Top..

Screening: a phenomenon known from electricity theory. As an example, the atom is made of an electrically charged nucleus and its geometric dimensions are very small. The nucleus is wrapped in a sort of electron cloud and in a non-ionized atom, the absolute value of the sum of the electron negative charges is equal to the nucleus' charge. At a relatively long distance from the atom, there is no electric field related to the atom in question, because the field of the electron charges cancels that of the nucleus charge (i.e., **screens** it). At a long distance from the atom, electrons totally "screen" the nucleus' field, but very close to the nucleus, they don't screen it at all; at some intermediate point there is a partial screening, which increases with the distance from the nucleus.

Schrödinger Equation: Quantum Mechanics' first successful equation. This is an approximative equation which does not deal with the spin and neglects Special Relativity Effects.

Spin: A property characterizing a quantum state in general and an elementary particle in particular. Allegorically, spin is often described as an angular momentum resulting from a particle's rotational movement around itself. The common measure unit of spins is actually ħ. The spin can have values of N/2, where N is an integral, non-negative number. Dirac particles have spin ½, and the photon has spin-1.

Strong Interaction: the force holding quarks together inside hadrons.

Strong Nuclear Force: a residual force of the Strong Interaction, binding protons and neutrons in the atomic nucleus.

Tau lepton, (sometimes called Tauon) is the third and heaviest electron-like particle. The Tau lepton is nearly 3500 times heavier than the electron and about 17 times heavier than the muon. Its half-life is measured in units of 10^{-13} seconds.

Van der Waals force: a force interacting between neutral molecules.

Wave: every particle has a wave quality, and therefore has a wavelength. A particle's wavelength depends on the particle's velocity (or, more accurately, on the particle's momentum). A fundamental wave property states that when a particle with a relatively large wavelength hits a target of composite particles, it does not notice the target's smaller constituents. Modern particle accelerators' growing capacity to accelerate particles to increasingly higher energy and momentum leads to the production of particle beams of increasingly smaller wavelengths, allowing researchers to obtain more detailed information on the target particles' structure. This is how, 100 years ago, the atomic structure had been studied, then some time later, the nucleus' structure was explored, and since about 40 years now, the proton's structure has been revealed. (Note: the medical imaging instrument called "ultra-sound scanner" is based on the same principle and uses an extremely short wavelength. The same is true for an electronic microscope where an electron with short wavelength enables detecting tiny details that cannot be seen by optical waves.)

Weak interaction: at energies that are not very high, the Weak Interaction is much weaker than the strong and the electromagnetic interactions. However, it can disintegrate particles and produce other particles. For example, a free neutron decays into a proton, plus an electron and an anti-neutrino as a result of the Weak interaction. At very high energies, such as those of the W particle, the weak interaction is quite significant and it is accountable of nearly one third of the disintegrations.

Selected Articles of Eliyahu Comay

During his scientific career Comay published theoretical physics articles in several fields, mainly nuclear physics, particle physics, quantum mechanics and classical electromagnetism. Below we include abstracts of his most important articles regarding his model of the strong interactions.

Axiomatic deduction of equations of motion in classical electrodynamics, Il Nuovo Cimento, vol 80B, N.2, 1984

The equations of motion of a classical system of electric charges, magnetic monopoles and electromagnetic waves are derived by using five axioms. The work answers the question: what are the equations of motion of this system which can be derived from a regular Lagrangian density?

Charges, monopoles and duality relations, Il Nuovo Cimento B, 1995, vol. 110, issue 11

A charge-monopole theory is derived from simple and self-evident postulates. Charges and monopoles take an analogous theoretical structure. It is proved that charges interact with free waves emitted from monopoles but not with the corresponding velocity fields. Analogous relations hold for monopole equations of motion. The system's equations of motion can be derived from a regular Lagrangian function.

Remarks on Photon-Hadron Interactions, Apeiron 10, No 2, 87 (2003)

Theoretical aspects of VMD and related approaches to real photon-hadron interaction are discussed. The work relies on special relativity, properties of linearly polarized photons, angular momentum conservation and relevant experiments. It is explained why VMD and similar approaches should not be regarded as part of a theory but, at most, as phenomenological models. A further experiment pertaining to this issue is suggested.

A regular theory of magnetic monopoles and its implications, published in "Has the Last Word been Said on Classical Electrodynamics?" (Rinton Press, NJ, 2004)

A regular charge-monopole theory is derived from simple and self-evident postulates. It is shown that this theory provides explanations for effects of strong and nuclear interactions. The theory is compared with Dirac's monopole theory. Applications to strong and nuclear interactions are compared with quantum chromodynamics. The results favor the regular charge-monopole theory and indicate difficulties of the other ones. An experiment that may provide further evidence helping to decide between the regular charge-monopole theory and quantum chromodynamics is suggested.

Difficulties with the Klein-Gordon Equation, Apeiron, Vol. 11, No. 3, 2004

Relying on the variational principle, it is proved that new contradictions emerge from an analysis of the Lagrangian density of the Klein-Gordon field: normalization problems arise and interaction with external electromagnetic fields cannot take place. By contrast, the Dirac equation is free of these problems. Other inconsistencies arise if the Klein-Gordon field is regarded as a classical field.

Further Difficulties with the Klein-Gordon Equation, Apeiron, Vol. 12, No. 1, 2005

Herein, the Dirac equation is compared with the Klein-Gordon equation. In contrast to the Dirac case, it is proved that the Klein-Gordon equation has difficulties with the Hamiltonian differential operator of relativistic quantum mechanics and with the definition of an inner product of wave functions, which is a requirement for a construction of a Hilbert space. An added discussion of the Pauli-Weisskopf article and that of Feshbach-Villars proves that their theories lack a self-consistent expression for the Hamiltonian. Related difficulties are pointed out.

The Yukawa Lagrangian Density is Inconsistent with the Hamiltonian, Apeiron, Vol. 14, No. 1, 2007

It is proved that no Hamiltonian exists for the real Klein-Gordon field used in the Yukawa interaction. It is also shown that a real Klein-Gordon particle can be neither in a free isolated state nor in a bound state having an angular momentum $l > 0$. The experimental data support these conclusions. This outcome is in a complete agreement with Dirac's negative opinion on the Klein-Gordon equation.

The Significance of Density in the Structure of Quantum Theories, Apeiron, Vol. 14, No. 2, 2007

It is proved that density plays a crucial role in the structure of quantum field theory. The Dirac and the Klein-Gordon equations are examined. The results prove that the Dirac equation is consistent with density related requirements whereas the Klein-Gordon equation fails to do that. Experimental data support these conclusions.

Remarks on the Proton Structure, Apeiron, Vol. 16, No. 1, 2009

Elastic and inelastic cross sections of proton-proton and electron-proton scattering are discussed. Special attention is given to elastic scattering and to the striking difference between the data of these two kinds of experiments. It is shown that the regular charge-monopole theory explains the main features of the data. Predictions of results of CERN's Large Hadron Collider are pointed out.

Physical Consequences of Mathematical Principles, Progress in Physics, 2009, Vol. 4

Physical consequences are derived from the following mathematical structures: the variational principle, Wigner's classifications of the irreducible representations of the Poincar´e group and the duality invariance of the homogeneous Maxwell equations. The analysis is carried out within the validity domain of special relativity. Hierarchical relations between physical theories are used. Some new results are pointed out together with their comparison with experimental data. It is also predicted that a genuine Higgs particle will not be detected.

Predictions of High Energy Experimental Results, Progress in Physics, Vol. 4, 2010

Eight predictions of high energy experimental results are presented. The predictions contain the Σ^+ charge radius and results of two kinds of experiments using energetic pionic beams. In addition, predictions of the failure to find the following objects are presented: glueballs, pentaquarks, Strange Quark Matter, magnetic monopoles searched by their direct interaction with charges and the Higgs boson. The first seven predictions rely on the Regular Charge-Monopole Theory and the last one relies on mathematical inconsistencies of the Higgs Lagrangian density.

On the Quantum Mechanical State of the Δ^{++} Baryon, Progress in Physics, Vol. 1, 2011

The Δ^{++} and the Ω^- baryons have been used as the original reason for the construction of the Quantum Chromodynamics theory of Strong Interactions. The present analysis relies on the multiconfiguration structure of states which are made of several Dirac particles. It is shown that this property, together with the very strong spin-dependent interactions of quarks provide an acceptable explanation for the states of these baryons and remove the classical reason for the invention of color within Quantum Chromodynamics. This explanation is supported by several examples that show a Quantum Chromodynamics' inconsistency with experimental results. The same arguments provide an explanation for the problem called the proton spin crisis.

Praise for Dr. Jane Murray and

Be the Change: Transforming Health Care from the Inside Out

"If Dr. Jane Murray ran things, we'd be a great deal healthier as a nation and when somebody did get sick, their care would be safer, more effective and far more compassionate that what the current system offers. Read her book. Then give a copy to your doctor and another to your Congressman."

-Victoria Moran, author of *Fit From Within* and *Creating a Charmed Life*

"What compelling reading with real innovations for revolutionizing health care. Your vision of healing hospitals is inspired – count me in! Congratulations on pulling all this together in a single volume, and for the courage that underlies it."

-Wyatt Townley, poet, yogi, and author of *Yoganetics* and *Breathing Lessons*

"This book is a tour-de-force – combining a wealth of personal experiences, relevant patient stories and clinical observations – all in the context of what you rightly call the long-standing crisis of health care in this country. You have drawn this all together in a lucid manner, in a pleasant straightforward style, firmly rooted in pertinent statistics and data. I was also struck by the recurring theme of what the environment for health care should look and feel like. We are enveloped in harsh and un-human places. Your own clinic exemplifies how a healing center can be calm, friendly and even serene. In trying to define what you have accomplished with this book, the term I came up with is 'Visionary Realism'. Congratulations!"

Fred Whitehead, author and philosopher

"I ordered your book and am so glad I did! It is fantastic! I find that I can open it to any page at random and glean some information I did not know – and that I totally agree with! Thanks for writing it!"

- Helen Wewers, civic leader and philanthropist